3D Information Technologies in Cultural Heritage Preservation and Popularisation

3D Information Technologies in Cultural Heritage Preservation and Popularisation

Editors

Marek Milosz
Jacek Kęsik

MDPI • Basel • Beijing • Wuhan • Barcelona • Belgrade • Manchester • Tokyo • Cluj • Tianjin

Editors

Marek Milosz	Jacek Kęsik
Lublin University of Technology	Lublin University of Technology
Poland	Poland

Editorial Office
MDPI
St. Alban-Anlage 66
4052 Basel, Switzerland

This is a reprint of articles from the Special Issue published online in the open access journal *Applied Sciences* (ISSN 2076-3417) (available at: https://www.mdpi.com/journal/applsci/special_issues/3D_information_technologies_in_cultural_heritage_preservation_and_popularisation).

For citation purposes, cite each article independently as indicated on the article page online and as indicated below:

LastName, A.A.; LastName, B.B.; LastName, C.C. Article Title. *Journal Name* **Year**, *Volume Number*, Page Range.

ISBN 978-3-0365-6281-0 (Hbk)
ISBN 978-3-0365-6282-7 (PDF)

Cover image courtesy of Jacek Kęsik

Contents

applied sciences

MDPI

Editorial

Special Issue on 3D Information Technologies in Cultural Heritage Preservation and Popularisation—Motivations, Works Overview, and the Future

Marek Milosz * and Jacek Kęsik *

Department of Computer Science, Lublin University of Technology, ul Nadbystrzycka 36B, 20-618 Lublin, Poland
* Correspondence: m.milosz@pollub.pl (M.M.); j.kesik@pollub.pl (J.K.)

Citation: Milosz, M.; Kęsik, J. Special Issue on 3D Information Technologies in Cultural Heritage Preservation and Popularisation—Motivations, Works Overview, and the Future. *Appl. Sci.* **2023**, *13*, 204. https://doi.org/10.3390/app13010204

Received: 7 December 2022
Accepted: 21 December 2022
Published: 23 December 2022

1. Introduction

Elements of cultural heritage preservation and popularisation are defined in the Convention Concerning the Protection of the World Cultural and Natural Heritage, which came into force in 1975 and obliges United Nations Educational, Scientific and Cultural Organization (UNESCO) member states to protect cultural and natural heritage [1]. This protection applies to the following areas: documentation, protection, reconstruction, restoration, conservation, dissemination, and popularisation [1].

Cultural heritage can be endangered by human actions (random or deliberate), forces of nature (floods, fires), or the mere passage of time. Examples of such activities, leading to the loss of elements of cultural heritage, are presented in [1], including such phenomena as [1]: the fire of the Notre-Dame Cathedral in Paris (France), the destruction of the Old Bridge in Mostar by tank fire (Bosnia and Herzegovina), the blowing up of the Temple of Bel in Palmyra (Syria), cannon shelling of Buddha statues in Bamiyan (Afghanistan), the fire of the National Museum in Rio de Janeiro (Brazil), the theft of petroglyphs from southwestern Arizona (USA), the destruction of Patan Durbar Square by earthquake (Nepal), and the natural collapse of the Azure Window in Gozo (Malta). Cases of irretrievable loss of elements of cultural heritage indicate the importance of its protection, including their documentation, but also popularisation.

Techniques for documenting the appearance of cultural heritage objects are currently being rapidly developed thanks to information technology (IT). Techniques such as sketching/photographing, manual measurement, or color photography are being replaced by modern IT technologies that enable the construction and use of 3D models.

At present, the following IT technologies are mainly used to obtain data on historical heritage cultural objects [1]:

(a) Photogrammetry;
(b) Scanning using structured light;
(c) Laser scanning.

These technologies are undergoing constant development, both in the field of hardware and software, and in the direction of automating information acquisition regarding objects of historical heritage. The monography [1] presents these technologies, including their parameters and exemplary data-processing software.

The dissemination of information on cultural heritage is now significantly simplified. Thanks to the development of IT, these activities are increasingly facilitated and can be easily implemented. The Internet has become a universal medium for delivering information to a mass audience, with its wide accessibility a consequence of near-universal global access to mobile telephony. As a consequence, the Internet has effectively and inexpensively popularized a very large part of the world's cultural heritage.

The topic of this Special Issue of the Applied Sciences journal "3D Information Technologies in Cultural Heritage Preservation and popularisation" is related to the interdisciplinary scope of scientific research, combining IT and the area of broadly understood culture. Interdisciplinary research directions in the area of science (i.e., computer science) combined with humanities (i.e., history, archaeology, museology, etc.) are very demanding in terms of carrying out scientific research and publishing results in scientific journals. Thus, a dedicated Special Issue meets the needs of this area of scientific interest.

2. Three-Dimensional Information Technologies in Cultural Heritage

In connection with the above, this Special Issue presents the results of research that applies the latest 3D information technologies to examine broadly understood cultural heritage. Eleven scientific articles have been published after a thorough and competent peer-review process.

The widest group of articles [2–7] are concerned with the 3D information technology techniques and methods that are used to protect and popularize the cultural heritage. Another group of papers [8–11] examine the practical aspects of using information technology to popularize cultural heritage and detect vandalism in the form of graffiti. Article [11], on the other hand, presents the methods used to select 3D scanners.

Neamtu et al. [2] present the results of the digital restoration and documentation of the Imperial Gates from the Wooden Church of Voivodeni in Sălaj County, Romania. The results of research on wooden churches from Transylvania in Romania, which are a unique and representative resource of rural cultural heritage in this region, are presented. Wooden churches are subject to natural degradation, especially as they are abandoned by their owners in favor of modern buildings. The research presented in this article combines XRF (X-ray fluorescence) and FTIR (Fourier transform infrared) spectroscopy, as well as augmented (AR) and virtual reality (VR) technologies.

Barszcz et al. compare two techniques of data acquisition and constructing 3D cultural heritage models [3]. They compare two object-scanning techniques: Structure-from-Motion (SfM) and Structured-light 3D Scanning (SLS). Their research consisted of photographing, using SfM, and scanning, using a special SLS device, the same historic object in the Afrasiyab museum in Samarkand, Uzbekistan. They describe data processing flows for both methods, and conclude that the cheaper and more widely available SfM method allows the construction of 3D models of objects of a sufficient quality to present and disseminate in media.

The article by Emmitt et al. [4] presents their research on a crested helmet from Tomb 1036 from the Casale del Fosso necropolis, Veii, Italy. In the research, the photogrammetry method was used to build a 3D mesh model of the object, as well as the X-ray fluorescence (XRF) method to analyze the helmet material's chemical composition. They visualize multiple XRF analysis results by constructing a 3D model of the helmet, which allows for a more intelligible presentation of data and helps in their interpretation.

The work by Żyła et al. [5] is a systematic review of 3D scanning techniques for historical costumes based on articles from eight databases, covering 2879 studies. The PRISMA protocol was adopted as the methodological basis for their literature research. Their basic conclusions indicate that a very small number of articles are related to the 3D scanning of historical costumes, the most commonly used techniques are photogrammetry and laser scanning, and the basic goals of most projects are to disseminate information about historical costumes.

Skublewska et al. present an innovative method of digitally capturing of folk dances [6]. They use a motion-capture technique and installation, as well as a developed wire model of the human hand, which complements the whole-body model. The above technique captures the movement of a dancer performing the Lazgi dance, which originates from the Khorezm region in Uzbekistan. Their research is one of only a few examples of digitizing intangible cultural heritage, such as dance or folk customs.

Wójcicki et al. use 3D scanning and an algorithm to highlight petroglyph patterns [7]. Their developed software analyzes 3D scans of petroglyphs in Hodjikent, Uzbekistan. They use the Structured-light 3D Scanning (SLS) method to acquire petroglyph data, and after appropriate post-processing, analyze the 3D model using an algorithm and a program developed by the authors. Despite their considerable results when highlighting petroglyph patterns, the authors stress the need to further refine the algorithm.

Merchan et al. [8] attempt to create a catalog of good practices in the use of augmented reality (AR) for the dissemination of rural architectural heritage. After an extensive literature review and based on their own experiences, the authors developed the Decalogue of Good Practices and present its implementation using the example of AR applications at the Archaeological Site "La Matilla" in Badajoz, Spain. They thoroughly describe the scheme of the procedure to produce the proof of concept of the AR application and illustrate its use with specific examples, emphasizing the problem of depopulation in some agricultural regions of Spain, as well as the importance of AR for their dissemination of information.

Comunità et al. [9] and Lim et al. [10] use IT to create 3D sound for cultural heritage. The first paper presents a set of web- and mobile-based applications for the curation and experience of 3D interactive soundscapes and sonic narratives in a cultural heritage context. They also present their methodology for evaluating the effectiveness of using the system to create a soundscape for cultural heritage preservation, including an analysis of soundscape curation and user experience. Their survey results indicate the practical usefulness of their developed platform. The second article (Lim et al. [10]) presents research related to establishing the role of social media in understanding of cultural diversity. A total of 22 representatives from cultural institutions participated in the study. Furthermore, they examined the participants' reception and impressions of a tour of central London, UK, which was supported by a narration emitted with 3D sound.

Choi et al. [11] focus on an automatic graffiti detection system using social networks, GPS, orthomaps, and artificial intelligence methods. The detector's effectiveness is presented using the Church of Agios Nikolaos (Leontariou), Kantza, Greece, as an example.

Beniak et al. explore the problem of selecting devices for the 3D scanning of cultural heritage objects [12], using an analytical hierarchical process (AHP) method to, for example, choose one of eight possible devices. They base their determination of the importance of experts for the assessment on four criteria. Finally, by analyzing the opinion of experts, they perform an objective device selection.

3. Future of 3D Information Technologies in Cultural Heritage

The call for papers for this Special Issue of the journal Applied Sciences has now closed. Despite this, the use of 3D information technology in cultural heritage requires further, increasingly advanced research. IT is becoming progressively more accessible and popular; therefore, contemporary culture and art must keep up with this trend. On the other hand, tools and software are constantly improving, enabling the automation of research and more accurate documentation of cultural heritage.

The editors of Applied Sciences considered this branch of science and IT applications and opened a new Special Issue titled: "3D Information Technologies for Tangible and Intangible Cultural Heritage" (https://www.mdpi.com/journal/applsci/special_issues/information_technologies_cultural_heritage, accessed on 2 December 2022).

Author Contributions: Conceptualization, M.M.; methodology, M.M.; validation, M.M. and J.K.; writing—original draft preparation, M.M.; writing—review and editing, J.K.; supervision, M.M. All authors have read and agreed to the published version of the manuscript.

Institutional Review Board Statement: Not applicable.

Informed Consent Statement: Not applicable.

Data Availability Statement: Not applicable.

Acknowledgments: We hereby thank all the authors and reviewers for their enormous and valuable contributions to this Special Issue. Thanks are due to all MDPI management and staff for their initiative and great help in completing this Special Issue, as well as congratulations for further enhancing the scientific impact of *Applied Sciences*.

Conflicts of Interest: The authors declare no conflict of interest.

References

1. Milosz, M.; Montusiewicz, J.; Kęsik, J. *3D Information Technology in the Protection and Popularisation of the Cultural Heritage of the Silk Road*; Politechnika Lubelska: Lublin, Poland, 2022; 134p. Available online: http://bc.pollub.pl/dlibra/publication/14114/edition/13771?language=en (accessed on 2 December 2022).
2. Neamțu, C.; Bratu, I.; Mărutoiu, C.; Mărutoiu, V.C.; Nemeș, O.F.; Comes, R.; Bodi, Ș.; Buna, Z.; Popescu, D. Component Materials, 3D Digital Restoration, and Documentation of the Imperial Gates from the Wooden Church of Voivodeni, Sălaj County, Romania. *Appl. Sci.* **2021**, *11*, 3422. [CrossRef]
3. Barszcz, M.; Montusiewicz, J.; Paśnikowska-Łukaszuk, M.; Sałamacha, A. Comparative Analysis of Digital Models of Objects of Cultural Heritage Obtained by the "3D SLS" and "SfM" Methods. *Appl. Sci.* **2021**, *11*, 5321. [CrossRef]
4. Emmitt, J.; McAlister, A.; Bawden, N.; Armstrong, J. XRF and 3D Modelling on a Composite Etruscan Helmet. *Appl. Sci.* **2021**, *11*, 8026. [CrossRef]
5. Żyła, K.; Kęsik, J.; Santos, F.; House, G. Scanning of Historical Clothes Using 3D Scanners: Comparison of Goals, Tools, and Methods. *Appl. Sci.* **2021**, *11*, 5588. [CrossRef]
6. Skublewska-Paszkowska, M.; Powroznik, P.; Smolka, J.; Milosz, M.; Lukasik, E.; Mukhamedova, D.; Milosz, E. Methodology of 3D Scanning of Intangible Cultural Heritage—The Example of Lazgi Dance. *Appl. Sci.* **2021**, *9*, 11568. [CrossRef]
7. Wojcicki, P.; Korga, S.; Milosz, M. Preliminary Application of the Algorithm Highlighting Petroglyph Patterns. *Appl. Sci.* **2022**, *12*, 1660. [CrossRef]
8. Merchán, M.J.; Merchán, P.; Pérez, E. Good Practices in the Use of Augmented Reality for the Dissemination of Architectural Heritage of Rural Areas. *Appl. Sci.* **2021**, *11*, 2055. [CrossRef]
9. Comunità, M.; Gerino, A.; Lim, V.; Picinali, L. Design and Evaluation of a Web- and Mobile-Based Binaural Audio Platform for Cultural Heritage. *Appl. Sci.* **2021**, *11*, 1540. [CrossRef]
10. Lim, V.; Khan, S.; Picinali, L. Towards a More Accessible Cultural Heritage: Challenges and Opportunities in Contextualisation Using 3D Sound Narratives. *Appl. Sci.* **2021**, *11*, 3336. [CrossRef]
11. Choi, J.; Toumanidis, L.; Yeum, C.M.; Charalampos, P.; Lenjani, A.; Liu, X.; Kasnesis, P.; Ortiz, R.; Jiang, N.-J.; Dyke, S.J. Automated Graffiti Detection: A Novel Approach to Maintaining Historical Architecture in Communities. *Appl. Sci.* **2022**, *12*, 2983. [CrossRef]
12. Beniak, J.; Šooš, Ľ.; Križan, P.; Matúš, M. Implementation of AHP Methodology for the Evaluation and Selection Process of a Reverse Engineering Scanning System. *Appl. Sci.* **2021**, *11*, 12050. [CrossRef]

applied sciences

MDPI

Article

Component Materials, 3D Digital Restoration, and Documentation of the Imperial Gates from the Wooden Church of Voivodeni, Sălaj County, Romania

Călin Neamţu [1,*], Ioan Bratu [2], Constantin Măruţoiu [3], Victor Constantin Măruţoiu [3], Olivia Florena Nemeş [3], Radu Comes [1], Ştefan Bodi [1,*], Zsolt Buna [1] and Daniela Popescu [1]

1 Department of Design Engineering and Robotics, Faculty of Machine Building, Technical University of Cluj-Napoca, 103-105 Muncii Blvd., 400641 Cluj-Napoca, Romania; radu.comes@muri.utcluj.ro (R.C.); zsolt.buna@muri.utcluj.ro (Z.B.); daniela.popescu@muri.utcluj.ro (D.P.)

2 National Institute for R&D of Isotopic and Molecular Technologies, 67-103 Donat St., 400293 Cluj-Napoca, Romania; ibratu@gmail.com

3 Faculty of Orthodox Theology, "Babeş-Bolyai" University, F/N Episcop Nicolae Ivan St., 400117 Cluj-Napoca, Romania; marutoiu.constantin@ot.ubbcluj.ro (C.M.); marutoiu.victor@ot.ubbcluj.ro (V.C.M.); marutoiu.olivia@ot.ubbcluj.ro (O.F.N.)

* Correspondence: calin.neamtu@muri.utcluj.ro (C.N.); stefan.bodi@muri.utcluj.ro (Ş.B.)

Citation: Neamţu, C.; Bratu, I.; Măruţoiu, C.; Măruţoiu, V.C.; Nemeş, O.F.; Comes, R.; Bodi, Ş.; Buna, Z.; Popescu, D. Component Materials, 3D Digital Restoration, and Documentation of the Imperial Gates from the Wooden Church of Voivodeni, Sălaj County, Romania. *Appl. Sci.* **2021**, *11*, 3422. https://doi.org/10.3390/app11083422

Academic Editor: Marek Milosz

Received: 15 March 2021
Accepted: 8 April 2021
Published: 11 April 2021

Abstract: The wooden churches from Transylvania, Romania, are a unique and representative cultural heritage asset for rural communities, both in terms of architecture and the style of painting that defines them as monuments of national heritage. These churches are in danger of degradation because rural communities are beginning to abandon them for various motives (e.g., they are too small, are expensive to maintain, or are being replaced by modern churches, built of stone and modern materials). The reason behind their accelerated degradation is that they are covered with shingles that need to be periodically changed and repaired to prevent water from reaching the inner painting layer, a process that is, in many cases, ignored. Imperial gates are the symbol of these churches and separate the nave from the narthex. They are made entirely out of wood and were sculpted and painted manually by skilled craftsmen and still represent the central element of these churches, in terms of art and aesthetics. The digital preservation of these heritage assets is an interdisciplinary undertaking, which begins with the physico-chemical analysis of the pigments in the painting layer, continues with three-dimensional (3D) digitization of the monument and of the objects of interest (such as the imperial gates), and finishes with a digital restoration of these monuments and artefacts. This paper presents a working methodology, successfully applied in digitizing and digitally restoring imperial gates from wooden churches in Transylvania, namely from the wooden church of Voivodeni, Sălaj County, Romania (Transylvania region). X-ray fluorescence and FTIR spectroscopy were used to determine the pigments in the painting layer of these artefacts, and after they were identified, they were synthesized in laboratory conditions. The resulting color was digitized and used for digitally restoring the artefact(s) to its (their) pristine condition. To popularize these cultural heritage assets, the authors make use of virtual reality to mediate the interaction between the general public and heritage objects in their current state of preservation, in a digital environment. Moreover, to showcase how these heritage objects were degraded over time, a digitally restored version of the artefact in pristine condition is presented alongside a version in its current state (as is, digitized, but not yet digitally restored).

Keywords: Fourier-transform IR spectroscopy; X-ray fluorescence spectroscopy; digital restoration of cultural heritage; digital preservation of imperial gates; spruce fir wood; painting materials; 3D scanning

1. Introduction

The central element of this paper is an interdisciplinary methodology, which employs state of the art tools, instruments, techniques, and equipment, found suitable by the authors, for capitalizing on a heritage asset (namely, the wooden church of Voivodeni, Sălaj County, Romania, (Figure 1), and the imperial gates from within (illustrated in Figure 2), through digital restoration/conservation and promotion to a wider audience.

In this regard, in the specialty literature we find a series of examples that use various digital technologies for the digital restoration and conservation of various heritage sites, such as: city centers [1], castles, or other monuments [2]. Furthermore, for studying painted artefacts, various techniques are used, such as: energy dispersive X-ray spectroscopy (EDX), polarized light microscopy (PLM), X-ray diffraction (XRD), micro-Raman spectroscopy (m-RS), Fourier-transform infrared spectroscopy (FTIR) and pyrolysis-gas chromatography-mass spectrometry (Py-GC/MS) [3]; in this field, the texture obtained by painters when using various brushes and painting techniques, is also studied [4].

This paper focuses on the "Holy Archangels Michael and Gabriel" wooden church from Voivodeni (Figure 1), which is situated on a hill nearby the village, as tradition dictated, and was constructed from dovetail joined wooden beams. The roof has steep angles, and the watchtower is placed on the roof ridge with the spire that surpasses the height of the walls. Inside the church lies an inscription documenting the year when construction started: "It was appointed for the church to be built up from the meadows in the year 1820. And it was built at the expense of the village, parish priest being Colciar Ștefan with his priest son Gheorghe Ștefan, and the curator was Elci Ioan. And the holy church was finished in the year 1822". The painter was the priest Iosif Perșe from Elciu, who painted this church and the imperial gates from within in the year 1832 [5–9].

The Four Evangelists are depicted on the two imperial wooden gates (Figure 2). Tradition assigned a symbol to each of them. Therefore, Saint Matthew is depicted together with a man (angel), Saint Marc is depicted together with a lion, Saint Luke is depicted together with a calf, and Saint John is joined by an eagle. The painting was deteriorated and needed immediate preservation and restoration.

Figure 1. The wooden church of Voivodeni, Sălaj County, Romania.

The investigation of the component materials (support materials and the painting layer) was completed through XRF and FTIR spectroscopic methods [10–16]. The sampling points for XRF and FTIR spectroscopy investigations are presented in Figure 2.

By employing such physico-chemical methods of investigation, the pigments used in the construction phase are identified and documented, because they represent the starting point in the classic restoration operations of any historical monument.

Furthermore, in order to contribute to the viability of the investigated monument, the authors propose a method of disseminating the results based on cutting edge technologies (namely, augmented and virtual reality—AR/VR). The popularization of this and other such monuments, based on this way of dissemination, may contribute to their reintroduction into religious and cultural tourism circuits, which, in term, would allow the collection and attraction of funds necessary for their maintenance and restoration (self-sustainment).

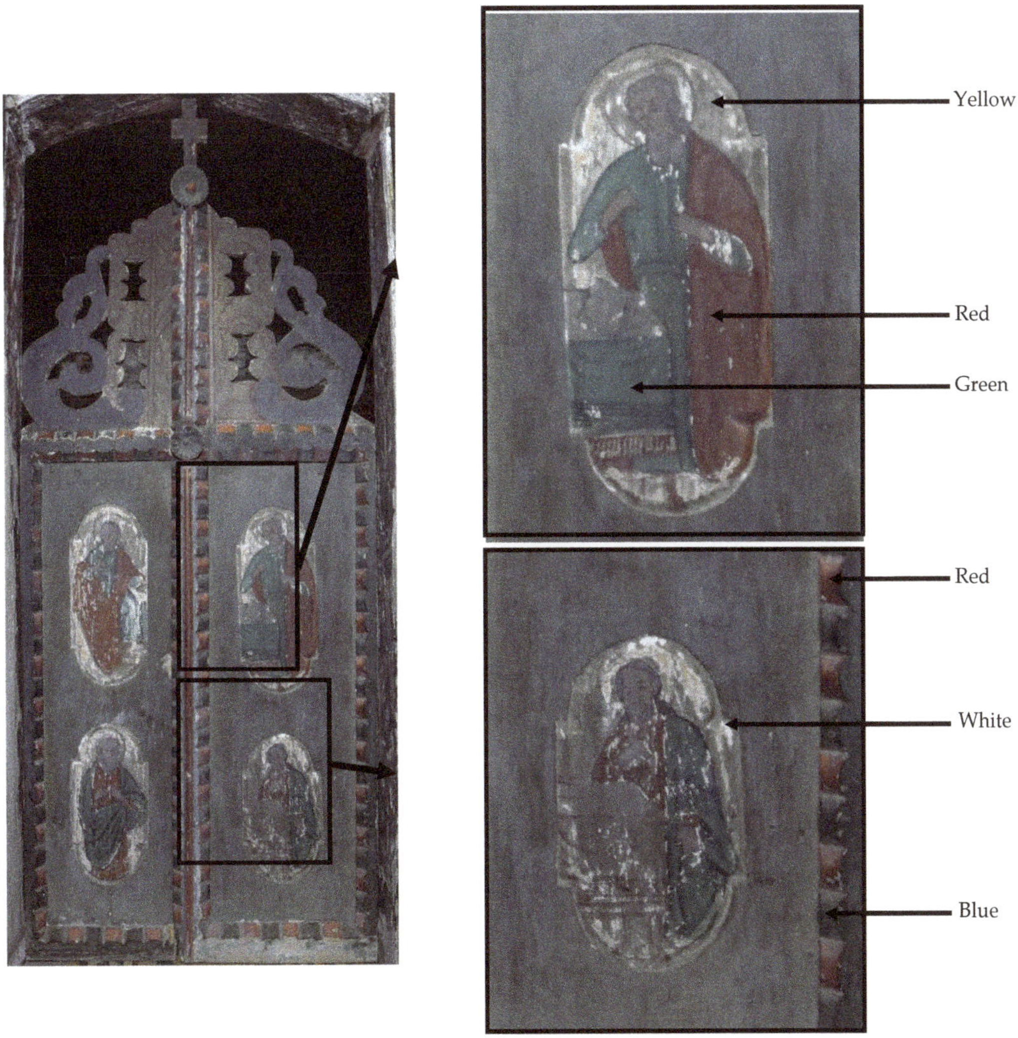

(a) (b)

Figure 2. The Imperial Gates of Voivodeni wooden church (**a**) and the indication of the sampling points (**b**).

2. Materials and Methods

The methodology (presented in Figure 3) for carrying out the tasks mentioned above (documentation, restoration, digital preservation, and dissemination) has been validated on several types of artefacts and cultural heritage objects, originating from the church, and it begins with the documentation from an artistic and historical point of view [15,17].

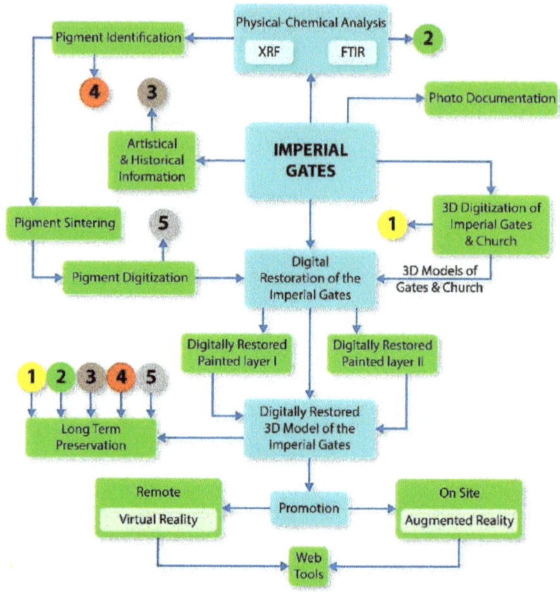

Figure 3. Methodology.

In the first part of the methodology, the artistic and historical documentation is made by art historians and aims to certify the heritage value of the investigated monument. This is completed by using bibliographic, online, and on-site sources.

The second stage consists of the photographic documentation of the monument; in this stage it is pursued the state-of-the-art documentation and conservation of the monument. Overview and detail high resolution photos are taken, in both visible and invisible spectrum (IR or UV), and primary information is collected for digitizing the monument.

The three-dimensional (3D) digitization of the monument can be carried out either by photogrammetry or 3D scanning, depending on required resolution, allotted time for scanning, level of detail, etc. The textured 3D model should present the current state of conservation as accurately as possible in order to document possible future restoration interventions.

The physico-chemical analyzes of the paint layer and the support materials should identify the composition of the pigments and help to identify the problems in the painting's support: pests, fungi, etc.

The identification of the pigments and their synthesis in the laboratory will determine the possible restoration operations, thus, it will be possible to choose compatible materials and it will be possible to study the effect of the current materials used in the restoration of the painting layer. After the pigment compositions are obtained, they can be digitized and documented for the digital preservation of the monument.

The digital restoration operation aims to improve the visual appearance of the monument in a virtual environment and to restore its appearance without any intervention (painted layer I) or its pristine appearance (painted layer 2—see) from the 1820s (when the church was built).

The digital restoration operation includes information from all stages described above: the scanned model is used to complete and remove the effects of time and other harmful factors on the monument, which is digitally repainted using digitized pigments obtained from the physico-chemical analysis. One can obtain two types of 3D models, one repainted with the initial colors and the other one painted with the current colors.

By going through the stages described so far, a data package is obtained that corresponds to the consolidated digital preservation model, and which contains the following information:

- Historical and artistic analysis;
- Complete photographic documentation of the monument;
- Documentation of the pigments and the state of conservation of the support of the pictorial layer;
- 3D model of the monument;
- Digitally restored 3D model.

After the completion of the consolidated data package, some elements of it can be further used for the online promotion of the monument both on-site (through AR applications or audio guided tours) and remotely (through VR-type applications—virtual tours).

The main goal of the previously described methodology is to preserve and digitally restore monuments in various stages of degradation. This approach is interdisciplinary, based on techniques from various fields of engineering, and is presented as follows.

2.1. X-ray Fluorescence

Non-destructive X-ray fluorescence elemental analysis (XRF) was performed with a handheld Bruker spectrometer (as shown in Figure 4), S1 TITAN series (EDXRF) configured with a Silicon diode PIN detector (SiPIN), Rh target X-ray tube with a maximum voltage of 50 kV. The use of this approach is validated in various studies, as well [18–20].

Figure 4. X-ray In-Situ investigation.

2.2. FTIR Spectroscopy

Fourier-transform Infrared (FTIR) measurements were performed with a Jasco 6100 spectrometer in the 4000 to 400 cm^{-1} spectral range with a resolution of 4 cm^{-1} (256 scans) by employing several (<1) mg of sample, and KBr pellet technique was applied. Similar investigations are detailed in other research endeavors, as well [18,21]. The obtained FTIR spectra were processed with the Spectral Analysis and the Origin 8.0 software. The samples were independently collected from the imperial gates and from mural paintings.

2.3. Digitization

The scanning and digitization of the imperial gates are aiming to preserve the conservation status in an electronic format. The 3D model is further used for various research activities and also for the promotion of these patrimony assets. The digitization was performed using a Creaform Go!SCAN 50 structured light scanner with a 50 μm dimensional precision able to acquire textures up to 268 Mpx. The digitization operation targeted both the shape and texture of the gates, the data being acquired simultaneously.

The 3D mesh resulted from scanning has approximately 4 million polygons/gate (Figure 5b), which were further transformed to a surface on which the texture (Figure 5a) was remapped. The Imperial Gate surface and the final 3D model are presented in Figure 6.

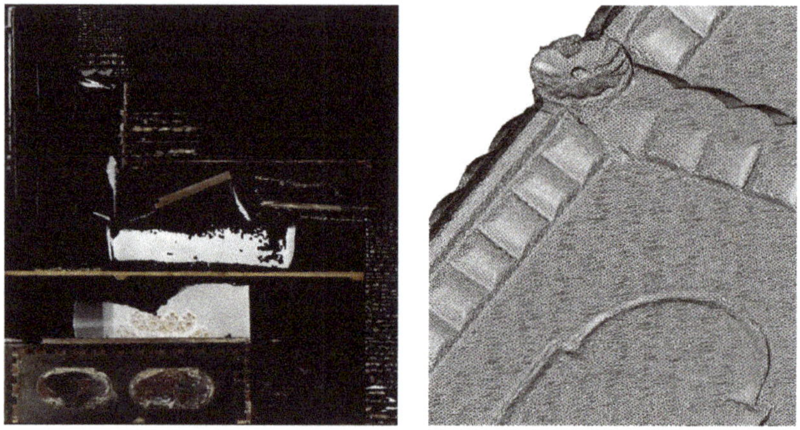

(a) (b)

Figure 5. The texture (**a**) and the mesh resulted from processing (**b**).

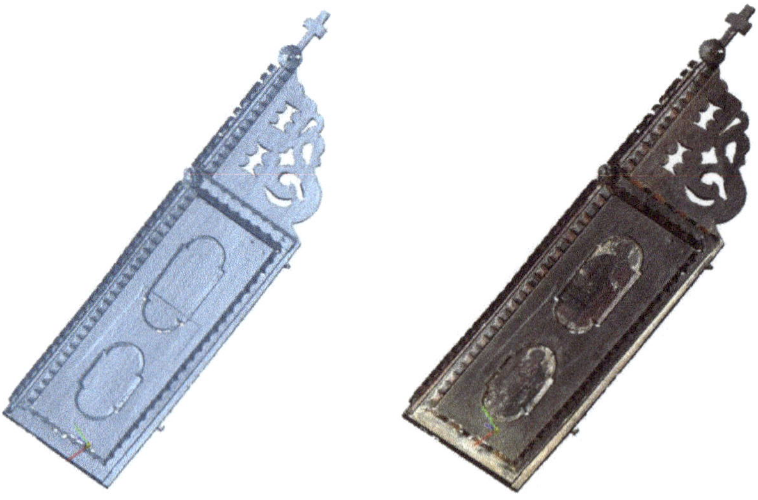

(a) (b)

Figure 6. The Imperial Gate's surface (**a**) and the final three-dimensional (3D) model (**b**).

3. Results and Discussion

In the following, the results obtained from deploying the methodology are presented, detailed, and applied by an interdisciplinary team of chemists, theologians, graphic artists, art historians, engineers, and IT specialists. The stages of physico-chemical analysis and digitization of the monument take place in parallel, the completion of the two stages condition the transition to the digital restoration stage.

3.1. Artistic and Historical Documentation

An element of these monuments, distinguished by its symbolic significance, are the imperial gates. These are essential components of the iconostasis of the Orthodox Christian churches, with symbolic value for cultural heritage. Byzantine iconography associates with the imperial gates the icon of the Annunciation, as the initial moment of saving humanity, through the symbolic opening of the gates of Heaven by Jesus.

In the Romanian Orthodox tradition, the transcendental passageway is made through three symbolic gates, through three levels of access: the entrance to the precinct, the access gate to the church and the imperial gates, accessible to priests. They separate the altar from the nave. Due to the size of the wooden churches from Transylvania, the imperial gates tend to take on greater importance in the eyes of the faithful.

Following the field study conducted between 2016 and 2019, it was found that the best preserved are those churches that are still functional. The depopulation of rural areas has reduced their number every year. Some of the inactive churches were saved by converting them to open-air village museums. Moreover, valuable artefacts were saved by moving them to these museums.

Of the 68 wooden churches from Sălaj County, Romania (including the three churches that were moved to open-air museums in Sibiu and Cluj-Napoca), only 21 preserve imperial gates, among which is also the church in Voivodeni.

In this stage, a photographic documentation is carried out, in both the visible and invisible spectrum (IR and UV), as shown in Figure 7. The purpose of this stage is to capture the current state of conservation (visible spectrum) and possible interventions or even helpful sketches used during the painting process by folk artists.

The photographic documentation is constituted in a document that will preserve in digital format the current state of conservation of the monument. DSLR cameras (Canon 5DSR—50Mpx) and a multispectral camera (modified Canon 550—18Mpx) were used to complete this documentation.

Figure 7. Photographs of Imperial Gates in the visible and IR spectrum.

In Figure 7, the first two images provide an overview of the whole artefact, while the last two showcase a close-up on the top right area.

3.2. Digitization

The wooden church has been scanned using a terrestrial laser scanner (Z + F 5010x) to obtain an accurate colored point cloud of the church's interior and exterior (Figure 8). The vast majority of digital reconstructions are based on 3D models created in various software, but the accuracy of a model generated on a computer, which is based on a layout sketch of the wooden church will not have the required accuracy in terms of geometrical dimensions. Therefore, most digital reconstruction models for cultural heritage assets are based on approximations. By using a state-of-the-art terrestrial laser scanner, it enables a true 1:1 scale digital recreation of the entire church (both interior and exterior) also the part involving the texturizing process is done automatically and accurately, since the scanner is capable to acquire the data as a colored point cloud (XYZRGB file extension).

Figure 8. Voivodeni wooden church point cloud and its surroundings, resulted from 3D scanning.

3.3. XRF

The purpose of the EDX-XRF spectroscopy was to determine the pigments from the painting materials; they are centralized in Table 1. The chemical elements of those pigments are detailed in Figure 9a–g.

Table 1. Sampled colors from the imperial gates and identified pigments (based on Figure 9).

Color	XRF Spectrum/Figure	Pigment
White	Figure 9a	Lead white
Yellow	Figure 9b	Orpiment
Ground	Figure 9c	Calcium salt based
Red	Figure 9d	Red mercury (cinnabar)
Blue	Figure 9e	Azurite
Green	Figure 9f	Malachite
Red	Figure 9g	Red iron

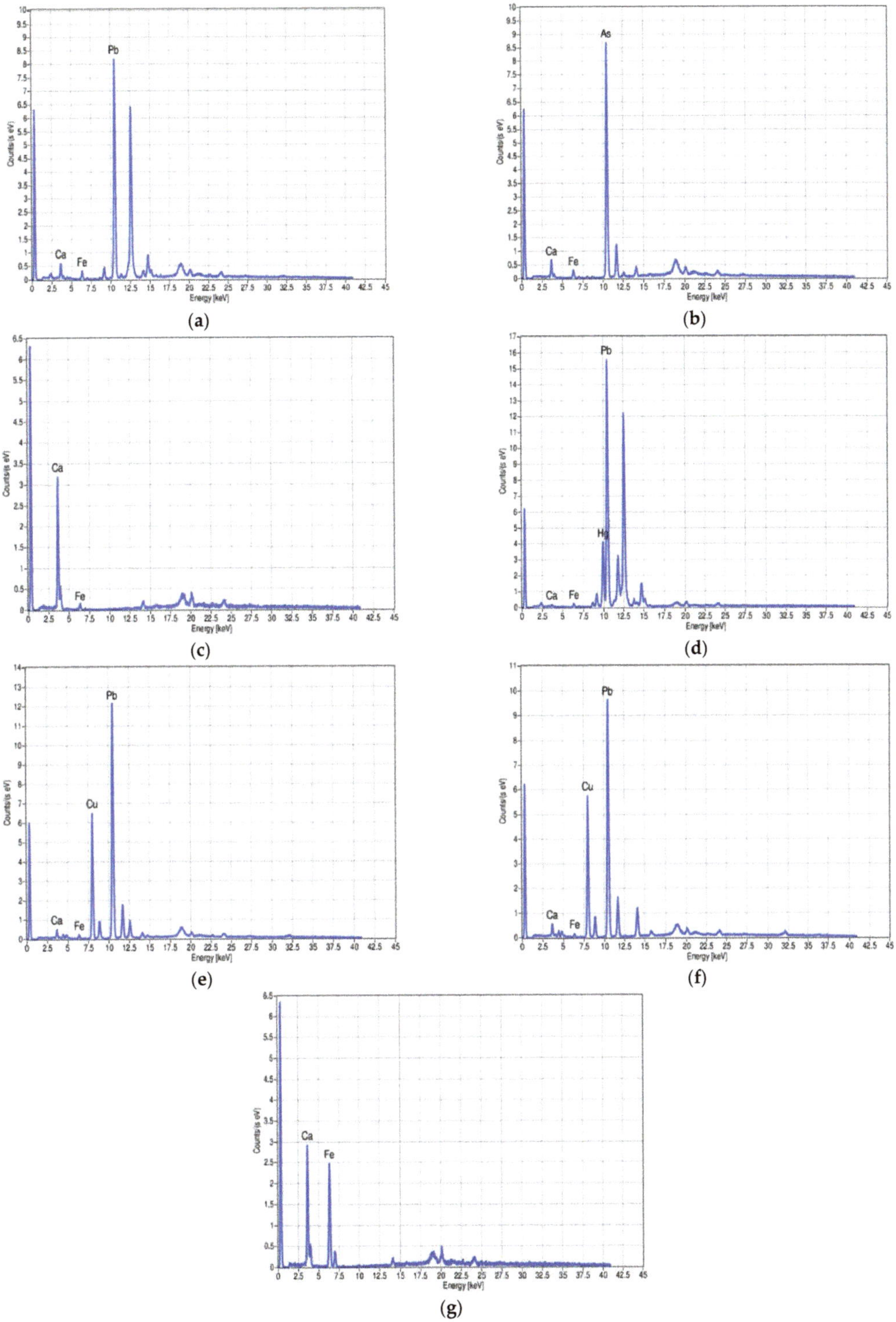

Figure 9. Chemical elements present in the painting materials.

3.4. FTIR Investigation of Imperial Gates Wood

From the comparison of FTIR spectrum of spruce fir with the FTIR spectrum of imperial gates (IG) wood, of the frame wood, see Figure 10, especially the 1100–900 cm^{-1} spectral range, spruce fir was identified as being employed for the IG wood and frame.

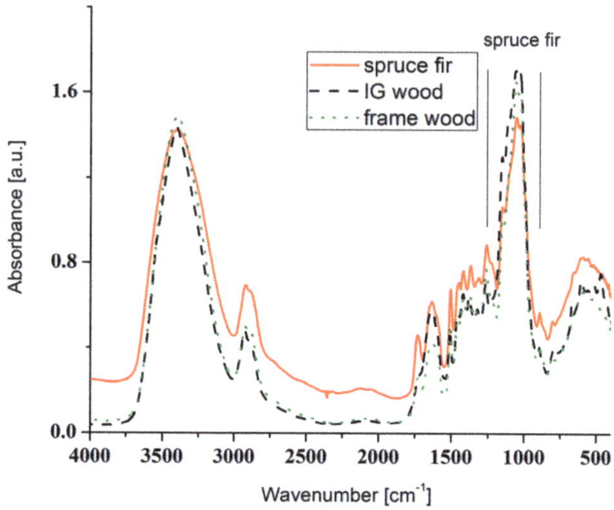

Figure 10. Imperial gates' wood- and frame wood FTIR spectra (as compared to the spruce fir one).

3.5. Painting Materials

The green painting material was identified as malachite and gypsum by the characteristic absorptions located at ~3600–3400 and 1200–1000 cm^{-1}, respectively, see Figure 11. Yellow painting material was identified by XRF as being orpiment. Egg yolk was identified by its specific absorptions present in the~3000–2800 and 1800–1500 cm^{-1} spectral domains (Figure 11). Red and Burgundy painting materials were identified as mixtures of red lead, red iron, and red mercury mixed with gypsum (Figure 12), based on XRF results.

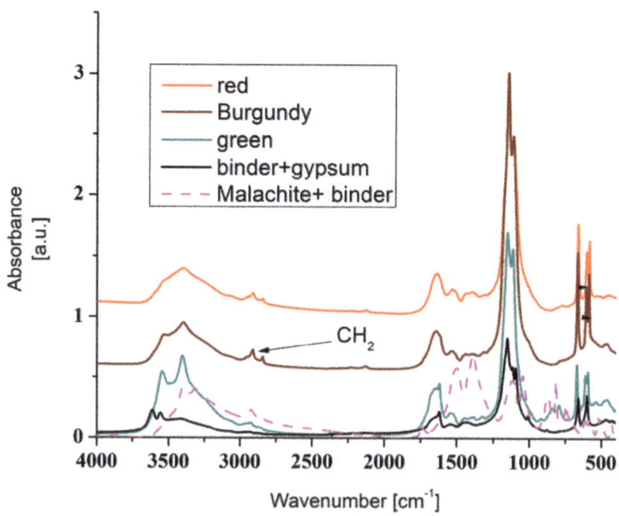

Figure 11. FTIR spectra of the painting materials, 4000 to 400 cm^{-1} spectral range.

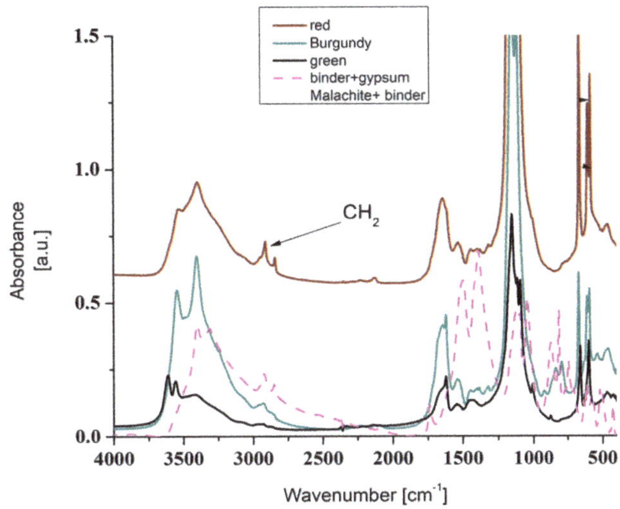

Figure 12. FTIR spectra of the painting materials, 4000 to 400 cm^{-1} spectral range, Oy extended.

3.6. Blue Pigment

Considering the data from XRF and FTIR, (Figures 13 and 14), one can identify traces of azurite or malachite mixed with gypsum, egg yolk, and linen oil in the blue painting material (identified absorptions at 3600–3400 cm^{-1} for gypsum, 2930 cm^{-1} for linen oil and egg yolk, 1647 cm^{-1} egg yolk protein, ~1100–1000 and ~615–580 cm^{-1} for gypsum).

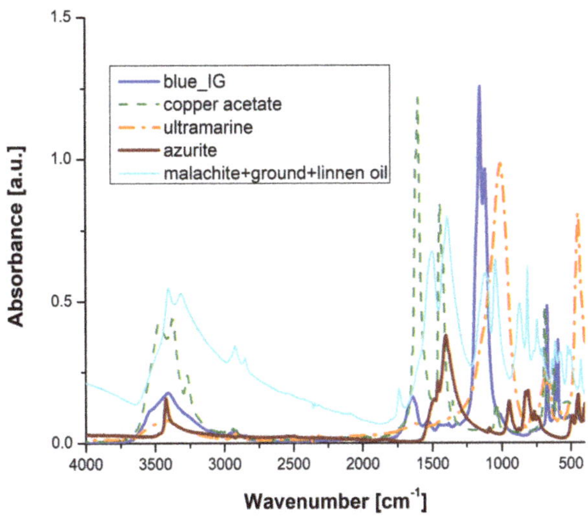

Figure 13. FTIR spectra of various blue and green painting materials, 4000 to 400 cm^{-1} spectral range.

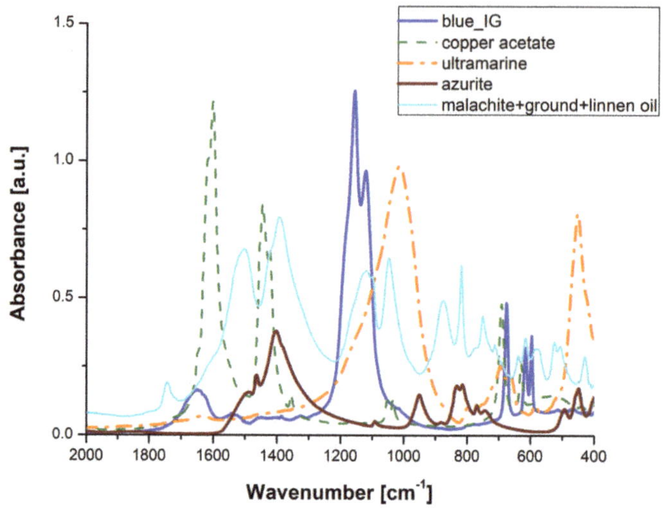

Figure 14. FTIR spectra of various blue and green painting materials, 2000 to 400 cm^{-1} spectral range, extended Oy axis.

4. Digital Restauration and VR Environment Integration of Imperial Gates

Digital conservation of the imperial gates is performed by storing the following data:

- Primary data resulted from scanning and photographing during scanning;
- Detailed 3D model resulted from processing data acquired by scanning without applying data reduction filters or texture compression;
- Optimized 3D model—this is a compressed model resulted from data reduction employing various filtering and compressing operations;
- All the physical and chemical tests performed on the painting layer and wooden support.

The detailed 3D model of the imperial gates (see Figure 6) has approximately 1.8 GB and it is difficult to be accessed by the public, being used mainly for research activities and digital restoration, as shown in [12]. Compressing using various methods [22–25] aims to obtain a simplified 3D model in order for it to be visualized and used on various cross-platform devices.

A virtual application was created in Unity software (version 2020.2.0f1) for the imperial gates, which facilitates their visualization employing HTC Vive. The user can move around these artefacts and can visualize them on a 1:1 scale.

Besides this application, the 3D model is shared on the sketchfab.com platform where users can manipulate the 3D model and stereoscopically visualize the imperial gates using a smartphone and a headset. Using the augmented reality facilities from this platform, the 3D models can be associated with a marker, visualized, and manipulated within the application (Figure 15). The 3D model of the gates can be visualized at https://skfb.ly/6ZXqx (published: 29 March 2021; accessed: 30 March 2021).

The process of preserving the current situation of the whole wooden church from Voivodeni was one of the most important objectives of the research, but our research is also focused on the potential use of digital image processing to recreate the paintings. As a case study, the digital restauration has been applied on one of the four evangelists, namely, on Saint Marc (upper part of Figures 2b and 16). The starting point for this step involves the use of the detailed 3D model, which had the precise 3D geometry as well as the mapped texture acquired using the hand-held Creaform Go!SCAN 50 structured light scanner. To obtain more details for the texture of each Saint, we have used a 50 Mpx full frame DSLR camera to gather a close-up of the painting, which was later merged into a single digital image. The merged high detailed image has been used as the starting point for the digital restoration process. To have a better lighting over the painting, the wooden gates were photographed outside in natural sunlight and artificial light, for a better exposure.

Figure 15. Showcasing the imperial gates using virtual reality.

The proposed digital reconstruction involved adjustment of the color saturation to better illustrate the initial situation and colors of the paintings. A wide variety of digital image manipulation has been applied to recolor the missing areas of the painting and adding details with various brush strokes, the proposed digital reconstruction is highlighted in Figure 16. However, here it must be noted that according to [26] virtual restoration is the process of "using a virtual model to reorder available material remains to visually recreate something that existed in the past. Thus, virtual restoration includes virtual anastylosis", which in turn "involves restructuring existing but dismembered parts in a virtual model". Moreover, because these techniques are deployed in the digital/virtual environment they do not interfere with the materiality of the original artwork, thus virtual restoration does not contravene with well-established principles of physical restoration, such as compatibility, reversibility, or minimal intervention [26,27]. Finally, the digitally restored model does not claim to have the same artistic value as the original, its purpose is to illustrate a probable variant of the original artwork, without the action of degrading factors upon it [28,29].

The proposed digital restoration has then been projected on the 3D model obtained using the 3D handheld structured light scanner. To generate a proposed digital restauration the whole right side of the imperial gate has also been recolored using the colors from the chromatic pallet of the digital image acquired by the structured light scanner. For obtaining the image from the right side of Figure 16, the colors used in the repainting process are obtained in laboratory conditions, based on the pigments identified as presented in Section 3.3, centralized in Table 1. By doing so, one can assume that the artefact is colored as its "pristine condition"—before it was affected by degrading (external) factors. The result of the recolored right side of the imperial gate is presented in the figure below, directly in a 3D texturing software (Substance Painter) that enables UV unwrapping and texturing.

In Figure 17, the 3D model and its material are presented, used for texturing purposes, in an unwrapped format; for the texture, a custom physically-based rendering (PBR)-type material was created, capable of reacting with the conditions from the virtual environment (lighting, shadows, reflections, etc.) and the mapping mode is "normal mapping", which is a technology of changing a normal pixel based on the chorochromatic map of normals, in which changing data are saved as texels. In this method, three texture channels are used on the map and, thanks to them, it has greater accuracy.

Figure 16. Proposed digital restoration of Saint Marc illustration on the imperial gates from Voivodeni.

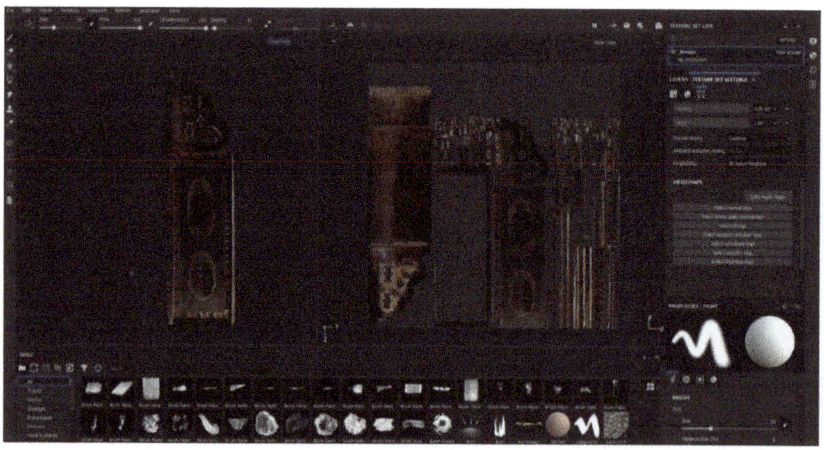

Figure 17. The recolored model of the right-side door of the imperial gates from Voivodeni.

The resulting 3D models of the imperial gates from Voivodeni have been included in a virtual reality application, which is dedicated to the imperial gates from Transylvania (Cluj and Sălaj counties, Romania). The VR application integrates multiple 3D scanned imperial gates and the application has been constantly updated with new scans. An older version of the application has been presented in an article focused on the multidisciplinary investigation of the imperial gate from Sălișca, Cluj county, Romania [15]. The application has been created using the Unity game engine software and it is compatible with a wide variety of head-mounted display (HMD) systems. The widespread adoption and compatibility with multiple VR systems have defined Unity as one of the most popular software solutions used to promote digital cultural heritage assets. The main advantage of transferring the 3D scanned model to virtual reality environments in comparison to other dissemination branches, such as webpage, is the ability to create interactive applications using a wide variety of sensors and controllers to manipulate or interact with the digital environment and allow users to fully immerse themselves (as showcased in this video: https://youtu.be/nnWSUrcDdZU—published: 8 March 2021; accessed: 9 March 2021).

For the promotion of the imperial gates, the virtual reality application was designed to work with HTC Vive/Cosmos and Valve Index virtual reality systems (HMD). The software behind the virtual reality tracking and interaction is based on SteamVR Software Development Kit (SDK) and it can work with various configurations—with HTC Vive wireless system up to Cosmos Elite—which does not require base to track the user's "virtual hands" as controllers.

The proposed virtual reality application integrates multiple imperial gates, which are paired with various metadata. On the walls behind each imperial gate, there are images and text metadata regarding the individual location. For some wooden churches, there are also videos that present the location of the wooden church and details regarding the 3D digitization of it.

The audio metadata files are positioned using a location bounding box; therefore, the user will be able to hear additional information regarding each individual wooden church and imperial gate, if he will be in the close proximity of those elements within the virtual environment. The figure below presents the location of the imperial gates from Voivodeni within the virtual environment. As can be observed, the left side of the imperial gates are the original models, as they have been obtained using 3D scanning, while for the ones in the right side, the digital reconstruction of Saint Marc is projected on top of the 3D scanned model, highlighting the difference between the original saturated colors of the painting and the current state.

Unfortunately, for the imperial gates from Voivodeni, they have not been physically curated and repainted, but hopefully there will be funding for this process as well as the whole painting from the interior of the church. Until then, the process of creating digital restoration is being used to raise awareness and to highlight the importance of restoring these types of cultural heritage assets that are present in most villages from Transylvania.

In Figure 18, to the left of the image, the imperial gates from Bulgari, Sălaj County, Romania, are presented, both in their initial state (to the right of the screen) and in their physical restored version (to the left of the screen). Both models have been 3D scanned by the team involved within this project using the same structured light scanner—Creaform Go!SCAN 50.

Another important feature of the virtual environment is based on the ability of the user to navigate the scene using the teleportation function. The system is capable of tracking the user in real time, but there are some limitations regarding the area where the users can move within the real environment, and the feature of teleportation helps to overcome this, easily. Teleportation can allow users to rapidly navigate the scene and to get close to the 3D scanned objects, allowing them to observe the detailed three-dimensional environment.

Using this feature, the users can easily look at the models from the other side and in some cases on the back of the imperial gates there are various elements as, for example, the imperial gates from Sânmihaiu Almașului, Sălaj County, have a text marking on the back, as well as the year 1816 (Figure 19).

The virtual reality application's camera tracking follows the position of the user, by default the camera is set in the middle of the virtual reality environment, where there is a table and three paintings. As soon as the user raises the HMD, the height of the camera will adjust to the user. Some of the imperial gates are above 2 m height, and some users, such as small children, may encounter problems when they try to spot various details on upper side of the gates; therefore, they can use the fly mode with the help of the controllers (by pressing two button—one for up and one for down), which enables them to ascend above the ground floor in the virtual reality environment. At a height of 4 m, there is a collision box, which keeps the user from elevating at a higher distance from the floor, keeping the users within the designated visualization environment.

A short video, illustrating the way in which a user interacts with the objects from the virtual environment, being fully immersed in it using an HMD, can be found at the following link: https://youtu.be/nnWSUrcDdZU (published: 8 March 2021; accessed: 9 March 2021).

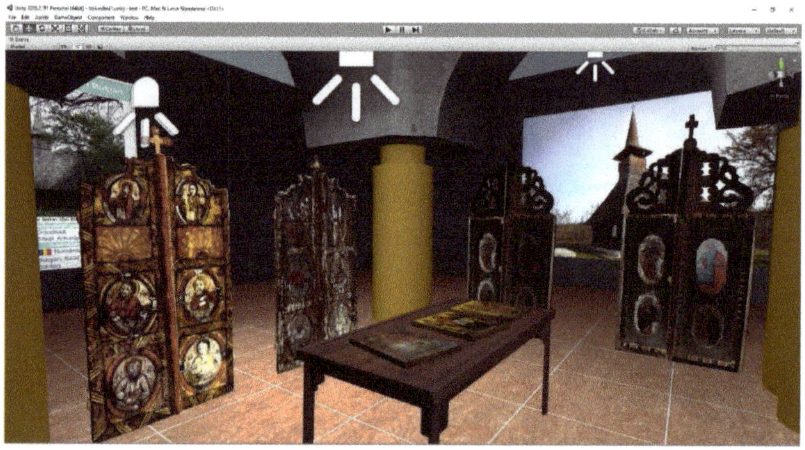

Figure 18. The imperial gates from Voivodeni positioned within the virtual reality environment.

Figure 19. The imperial gates from Sânmihaiu Almașului, Sălaj County, back details.

5. Conclusions

Using an interdisciplinary approach, based on physico-chemical analysis and 3D digitization techniques, long-term digital preservation is possible with an information package, which provides an overview and accurate data about the imperial gates. After identifying and documenting the heritage asset, the interdisciplinary team carried out a massive in-situ data collection, which includes physico-chemical analyzes for each color (performed in different sampling places), 3D digitization of the monument and of the imperial gates, and artistic and historical documentation.

The physico-chemical analyzes revealed that the woods used to craft the imperial gates and their frame are spruce fir and the component materials used in the painting process are ground gypsum; binder—egg yolk; red painting material—red iron, red lead, red mercury; yellow—orpiment; green—malachite; blue—azurite; white—lead white; egg yolk painting technique.

All of these results can further be used in the restoration of the entire monument and, at the same time, they are digitally preserved and stored together with all of the artistic and historical information, with the 3D model of the church and of the imperial gates, in their current state.

The digital restoration of the imperial gates can be carried out using the pigments identified by the physico-chemical analyzes, applied on the 3D model of the church and the imperial gates. The later ones can be repainted with their current colors (painted layer I) or with the initial colors (from their pristine state), obtained by synthesizing them in the laboratory, based on the pigments identified in the painted layer (painted layer II).

The 3D models can be used in virtual reality applications, which allow access to a much larger number of potential stakeholders, than in-situ visiting, and in the particular case of the imperial gates, allows the general public to see also the backside of the gates, which within the church is forbidden to female persons. Increasing the number of users who can examine the models and interact with them, the "appetite" of visitors for such cultural assets can be stimulated; thus, their reintroduction into the cultural and religious tourist circuits can be achieved.

Finally, it is worth mentioning that the methodology presented in this paper was validated by the authors on five different imperial gates, with various particularities (such as the sculpting complexity—Figure 19—or with multiple painting layers—) being presented in the developed VR application (its fundamentals and practicalities were detailed in the previous section).

Author Contributions: Conceptualization, C.N., I.B., C.M. and D.P.; methodology, C.N., D.P., R.C., Ș.B. and Z.B.; software, V.C.M., O.F.N., R.C., Ș.B., D.P. and Z.B.; validation, C.N., V.C.M. and O.F.N.; formal analysis, V.C.M., O.F.N., R.C., Ș.B., D.P. and Z.B.; investigation, V.C.M. and O.F.N.; resources, C.N., D.P., R.C. and Ș.B.; data curation, V.C.M. and O.F.N.; writing—original draft preparation, C.N., V.C.M., Ș.B. and R.C.; writing—review and editing, C.N., Ș.B. and R.C.; visualization, C.N., R.C., Ș.B. and Z.B.; supervision, C.N., I.B., C.M. and D.P.; project administration, C.N., I.B., C.M. and D.P.; funding acquisition, C.N. and D.P. All authors have read and agreed to the published version of the manuscript.

Funding: This work was supported by a grant from the Romanian Ministry of Research and Innovation, CCCDI—UEFISCDI, project number PN-III-P1-1.2-PCCDI-2017-0812/53PCCDI, within PNCDI III.

Institutional Review Board Statement: Not applicable.

Informed Consent Statement: Not applicable.

Data Availability Statement: Not applicable.

Conflicts of Interest: The authors declare no conflict of interest.

References

1. Rossato, L.; Massai, P.; Maietti, F.; Balzani, M. Digital Tools for Documentation and Analysis of Vernacular Cultural Heritage in Indian City Centers. *Int. J. Archit. Herit.* **2019**. [CrossRef]
2. García-Molina, D.F.; González-Merino, R.; Rodero-Pérez, J.; Carrasco-Hurtado, B. 3D Documentation for the Conservation of Historical Heritage: The Castle of Priego de Córdoba (Spain). *Virtual Archaeol. Rev.* **2020**, *12*, 115–130. [CrossRef]
3. Peng, F.; Ge-Le, T.; Jing, L.; Jia-Xin, L.; Yu-Hu, L.; Hong, Y. Investigation of Ancient Architectural Painting from the Taidong Tomb in the Western Qing Tombs, Hebei, China. *Coatings* **2020**, *10*, 688. [CrossRef]
4. Mironova, A.; Robache, F.; Deltombe, R.; Guibert, R.; Nys, L.; Bigerelle, M. Digital Cultural Heritage Preservation in Art Painting: A Surface Roughness Approach to the Brush Strokes. *Sensors* **2020**, *20*, 6269. [CrossRef]

5. Cristache-Panait, I. *Biserica Sf. Arhangheli Mihail și Gavril din Voivodeni. Monumente istorice bisericești din Eparhia Ortodoxă Română a Oradei. Biserici de lemn*; The Romanian Orthodox Episcopy of Oradea: Oradea, Romania, 1978; pp. 430–432.
6. Cristache-Panait, I. Bisericile de lemn din Sălaj. *Buletinul Monumentelor Istorice* **1971**, *1*, 31–40.
7. Ghergariu, L. *Biserici de lemn din Sălaj*; MSS in National Archives of Zalău, Personal Collection of Ghergariu Leontin, act 11 of 1976; National Archives of Zalău: Zalău, Romania, 1976.
8. Măruțoiu, C.; Popescu, D.; Bratu, I.; Moraru, R.; Postolache, D.L. *Biserici de lemn din Transilvania*; Mega: Cluj-Napoca, Romania, 2017; ISBN 978-606-543-857-6.
9. Drăguţ, V. *Dicţionar Enciclopedic de Artă Medievală Românească*; Editura Științifică și Enciclopedică: Bucharest, Romania, 1976; pp. 35–36.
10. Nemeș, O.F.; Bratu, I.; Măruțoiu, C.; Postolache, D.L.; Dragomir, M.; Tănăselia, C.; Garabagiu, S. Scientific Investigations of the Imparial Gates from Petrindu Wooden Church, Sălaj County, Romania. *X-ray Spectrom.* **2018**, *47*, 176–185.
11. Măruțoiu, C.; Bratu, I.; Nemeș, O.F.; Dit, I.-I.; Comes, R.; Tănăselia, C.; Fălămaș, A.; Miclăus, M.; Marutoiu, V.C.; Moraru, R. Instrumental Analysis of Materials and Topology of the Imperial Gates Belonging to the Apahida Wooden Church, Cluj County. *Vibr. Spectrosc.* **2017**, *89*, 131–136. [CrossRef]
12. Bratu, I.; Siluan, M.; Măruțoiu, C.; Kacso, I.; Garabagiu, S.; Măruțoiu, V.C.; Tănăselia, C.; Popescu, D.; Postolache, D.L.; Pop, D. Science Applied for the Investigation of Imperial Gate from Eighteenth Century Wooden Church of Nicula Monastery. *J. Spectrosc.* **2017**. [CrossRef]
13. Moldovan, Z.; Bratu, I.; Măruțoiu, C.; Kacso, I.; Trosan, L.; Pop, D.; Nemeș, O.F.; Tănăselia, C. Characterization of an Eighteenth-Century Wooden Icon from the Ethnographic Museum of Transylvania. *Anal. Lett.* **2016**, *49*, 2597–2605. [CrossRef]
14. Hernanz, A.; Bratu, I.; Măruțoiu, O.F.; Măruțoiu, C.; Gavira-Vallejo, J.M.; Edwards, H.G.M. Micro-Raman Spectroscopic Investigation of External Paintings from St. Dumitru's Church, Suceava, Romania. *Anal. Bioanal. Chem.* **2008**, *392*, 263–268. [CrossRef] [PubMed]
15. Neamțu, C.; Măruțoiu, V.C.; Bratu, I.; Măruțoiu, O.F.; Măruțoiu, C.; Chirilă, I.; Dragomir, M.; Popescu, D. Multidisciplinary Investigation of the Imperial Gates Belonging to the 17th Century Wooden Church of Sălișca, Cluj County, Romania. *Sustainability* **2018**, *10*, 1503. [CrossRef]
16. Gomes, L.; Bellon, O.R.P.; Silva, P. 3D reconstruction methods for digital preservation of cultural heritage: A survey. *Pattern Recognit. Lett.* **2014**, *50*, 3–14. [CrossRef]
17. Măruțoiu, C.; Bratu, I.; Nemeș, O.; Nemeș, D.; Neamțu, C.; Moldovan, Z.; Tia, T.; Udrea, I.; Tigae, C. Scientific investigation of the paintings from the Agârbiciu (Cluj County) Wooden Church. *J. Min. Mat. Charact. Eng.* **2020**, *8*, 177–196. [CrossRef]
18. Zamboni, C.B.; Redígolo, M.M.; Miura, V.T.; Costa, I.; Nagai, M.L.E.; Salvador, P.A.V.; Nogueira da Silva, D.G. Non-Destructive Analysis in the Study of Historical Photographs by pXRF and ATR-FTIR spectroscopies. *J. Forensic Sci.* **2021**. [CrossRef]
19. Molari, R.; Appoloni, C.R.; Hennen Rodriguez, S. Non-Destructive Portable X-ray Fluorescence Analysis of the Portrait of a Young Man with a Golden Chain (c. 1635) by Rembrandt and/or Atelier. *Appl. Radiat. Isot.* **2020**, *165*. [CrossRef]
20. Hennen Rodriguez, S.; Appoloni, C.R.; Campos, P.H.O.V.; Gonçalves, B.; Kajiya, E.A.M.; Molari, R.; Rizzutto, M.A.; Winter, C. Non-Destructive and Portable Analyses Helping the Study and Conservation of a Saraceni Copper Plate Painting in the São Paulo Museum of Art. *Microchem. J.* **2020**, 155. [CrossRef]
21. Retko, K.; Legan, L.; Ropret, P. SERS Procedure Using Photoreduced Substrates and Reflection FTIR Spectroscopy for the Study of Natural Organic Colourants. *J. Raman Spectrosc.* **2021**, *52*, 130–144. [CrossRef]
22. Neamțu, C.; Popescu, D.; Mateescu, R.; Suciu, L.; Hurgoiu, D. About Quality and Properties of Digital Artifacts. *Mediterr. Archaeol. Archaeom.* **2014**, *14*, 55–64.
23. Hatzopoulos, J.N.; Stefanakis, D.; Georgopoulos, A.; Tapinaki, S.; Pantelis, V.; Liritzis, I. Use of Various Surveying Technologies to 3D Digital Mapping and Modelling of Cultural Heritage Structures for Maintenance and Restoration Purposes: The Tholos in Delphi, Greece. *Mediterr. Archaeol. Archaeom.* **2017**, *17*, 311–336. [CrossRef]
24. Comes, R.; Tompa, V.; Bodi, Ș.; Neamțu, C.; Torok, K.F. From Theory to Practice: Digital Reconstruction and Virtual Reality in Archaeology. *J. Anc. Hist. Arch.* **2018**, *4*, 80–88. [CrossRef]
25. Jamhawi, M.; Alshawabkeh, Y.; Freewan, A.; Al-Gharaibeh, R. Combined Laser Scanner and Dense Stereo Matching Techniques for 3D Modelling of Heritage Sites: Dar Es-Saraya Museum. *Mediterr. Archaeol. Archaeom.* **2016**, *16*, 185–192. [CrossRef]
26. Lopez-Menchero, V.M.; Grande, A. The Principles of the Seville Charter. In Proceedings of the XXIInd CIPA Symposium, Prague, Czech Republic, 12–16 September 2011.
27. Xu, J.; Ding, L.; Love, P.E.D. Digital Reproduction of Historical Building Ornamental Components: From 3D Scanning to 3D Printing. *Autom. Constr.* **2017**, *76*, 85–96. [CrossRef]
28. Vico, L. Authenticity and Realism: Virtual vs. Physical Restoration. In *Authenticity and Cultural Heritage in the Age of 3D Digital Reproductions*; Di Giuseppantonio Di Franco, P., Galeazzi, F., Vassallo, V., Eds.; McDonald Institute: Cambridge, UK, 2018; pp. 25–33. [CrossRef]
29. Limoncelli, M. *Virtual Hıerapolıs. Virtual Archaeology and Restoration Project (2007–2015), Hierapolis di Frigia XIII*; Ege Yayinlari: Istanbul, Turkey, 2019; ISBN 978-605-7673-10-7.

applied sciences

MDPI

Article

Comparative Analysis of Digital Models of Objects of Cultural Heritage Obtained by the "3D SLS" and "SfM" Methods

Marcin Barszcz [1,*], Jerzy Montusiewicz [1], Magdalena Paśnikowska-Łukaszuk [2] and Anna Sałamacha [1]

[1] Department of Computer Science, Lublin University of Technology, Nadbystrzycka 36B, 20-618 Lublin, Poland; montusiewicz@pollub.pl (J.M.); a.salamacha@pollub.pl (A.S.)
[2] Department of Fundamentals of Technology, Lublin University of Technology, Nadbystrzycka 38, 20-618 Lublin, Poland; m.pasnikowska-lukaszuk@pollub.pl
* Correspondence: m.barszcz@pollub.pl

Abstract: In the era of the global pandemic caused by the COVID-19 virus, 3D digitisation of selected museum artefacts is becoming more and more frequent practice, but the vast majority is performed by specialised teams. The paper presents the results of comparative studies of 3D digital models of the same museum artefacts from the Silk Road area generated by two completely different technologies: Structure from Motion (SfM)—a method belonging to the so-called low-cost technologies—and by Structured-light 3D Scanning (3D SLS). Moreover, procedural differences in data acquisition and their processing to generate three-dimensional models are presented. Models built using a point cloud were created from data collected in the Afrasiyab museum in Samarkand (Uzbekistan) during "The 1st Scientific Expedition of the Lublin University of Technology to Central Asia" in 2017. Photos for creating 3D models in SfM technology were taken during a virtual expedition carried out under the "3D Digital Silk Road" program in 2021. The obtained results show that the quality of the 3D models generated with SfM differs from the models from the technology (3D SLS), but they may be placed in the galleries of the vitrual museum. The obtained models from SfM do not have information about their size, which means that they are not fully suitable for archiving purposes of cultural heritage, unlike the models from SLS.

Keywords: structure from motion (SfM); structured-light 3D scanning (3D SLS); 3D models; cultural heritage; Silk Road; Afrasiyab museum; virtual museum

Citation: Barszcz, M.; Montusiewicz, J.; Paśnikowska-Łukaszuk, M.; Sałamacha, A. Comparative Analysis of Digital Models of Objects of Cultural Heritage Obtained by the "3D SLS" and "SfM" Methods. *Appl. Sci.* **2021**, *11*, 5321. https://doi.org/10.3390/app11125321

Academic Editor: Theodore E. Matikas

Received: 15 May 2021
Accepted: 5 June 2021
Published: 8 June 2021

1. Introduction

3D digitisation of museum resources is becoming a more and more frequent practice, but the vast majority is performed by specialised teams representing the scientific community or professional companies offering such services. In recent years, data acquisition technologies have been developed to create three-dimensional models that use data belonging to various types of information. In photogrammetry [1] and Structure from Motion (SfM) [2], a collection of photos is used to build 3D models, which in successive shots of an object cover at least 60% of the surface. Laser scanning or structured-light 3D scanning (3D SLS) provides a cloud of points (x, y coordinates from the object's surface), which ultimately allows us to build its 3D model. In addition to the acquisition of surface point coordinates, colour data are also collected, thanks to which the generated digital model obtains realistic features. In the era of the global pandemic caused by the COVID-19 [3] virus, when excursion and pilgrimage traffic has been halted or significantly restricted, many cultural heritage museum institutions wish to maintain their presence in the tourism market by introducing traditional exhibitions into digital space. The attractiveness of such digital exhibitions, increasingly known as Virtual Museum (VM), will largely depend not only on the uniqueness of the exhibits presented, but also on their digital quality. SfM technology belongs to the so-called low-cost technologies, which may also suggest that the models obtained in this way will not be of sufficiently good quality. The paper presents the results of comparative studies of digital models of the same museum artefacts from the Silk Road area generated by two completely different technologies: SfM and 3D SLS. The developed three-dimensional models based on the point cloud obtained from the scanning were created from data collected in the Afrasiyab museum in Samarkand (Uzbekistan) during "The 1st Scientific Expedition of the Lublin University of Technology to Central Asia" in 2017 [4]. Photos for creating 3D models in SfM technology were taken during a virtual expedition carried out as part of the "3D Digital Silk Road" [5] program in 2021.

The purpose of the article can be summarized as follows:

- Comparison of data acquisition procedures for artefacts of material cultural heritage in SfM and 3D SLS technologies.
- Generating 3D objects of museum artefacts in SfM and 3D SLS technology.
- Assessment of parameters and quality of 3D models obtained in SfM and 3D SLS technologies.

2. Background Study

In the last two decades, the acquisition of three-dimensional data by laser scanning [6], using structured light [7] or SfM [2,8] has become one of the standard methods of documenting cultural heritage. It makes it possible to analyse the shape, geometry historic artefacts in a safe way [9]. They do not damage the surface or structure of the exhibits, because the activities carried out using the abovementioned methods do not require direct contact with the analyzed object. They allow for the accurate development of digital documentation of the measured object and then, for example, for creating a copy using the 3D printing technique [10].

The interest in scanning techniques is constantly growing and is being developed [11], because digitisation of museum collections allows to preserve exhibits for many decades and to pass on modern scientific achievements to the next generations. Creative involvement in digital 3D models of heritage artefacts can stimulate learning and support new forms of activity using digital models of historic artifacts [12], e.g., it allows disseminating 3D models on the Internet or creating virtual museums. Highly accurate documentation and 3D reconstructions are of fundamental importance for analysis and further interpretations in archeology. In recent years, integrated digital research (ground-based research methods and UAV photogrammetry) has proven its major role in documenting and understanding excavation contexts. This has become possible thanks to the development of specialised devices and the development of appropriate methods for collecting a wide variety of data on site. Several historic sites in the Sepino area of Italy were investigated with these methods and were realised using both range-based and image-based data acquisition methods. The laser scanning data were combined with SfM clouds in the same frame of reference, determined by topographic and GPS measurements [13]. By means of photogrammetry, the area of the Great Plains was examined under field conditions [14], as well as areas of Jordan [15]. The study of the effects of deformation in thermally forced rock masses has aroused growing interest in engineering geologists in the last decade. In this context, digital photogrammetry and infrared thermography and laser scanning have become the two most frequently used analysis techniques in engineering geology applications as they can provide useful information on the geomechanical and thermal conditions of these complex natural systems [16,17]. Photogrammetry and 3D scanning were used in the Pinchango Alto research in Peru [18]. 3D scanning methods were also discussed in the [19] work on the virtual archaeological exhibition. In the study of a unique early Christian carved stone discovered in Scotland, 2D photos were used as the basis for the reconstruction of the 3D model using the SfM method [20]. A FARO Photon Laser Scanner 120/20 system was used to record the geometric data of a complex of historic buildings in Malacca (Malaysia), declared a UNESCO World Heritage Site in 2008 [21]. Scanning with the use of 3D technology was also used to create ethnographic collections [22].

Having digital 3D models allows one to perform a virtual reconstruction of incomplete or damaged artefacts by restoring their original shapes or decorations along with the functions they performed. An example is the reconstruction of a Roman cornice from the Castulo archaeological site (located in Spain) presented in [23]. Virtual 3D models were created on the basis of a series of photos in the paid Agisoft Metashape program, and digital reconstruction using the free Blender software. The reconstructed models were then used to recreate the cavities by 3D printing. Thanks to digital reconstruction, exhibitions in virtual museums gain a new format. Examples of virtual museums created with the use of 3D models obtained in the process of digitisation of objects are the Archaeological Museum in Zagreb [24] and the National Archaeological Museum in Florence [25]. Artefact scans are described in [26], and the obtained models were used to create integrated VR applications.

SfM technology is sometimes dedicated to specific devices. The work [27] describes an example of 3D digitisation of a dress using specialised Autodesk 123D Catch software developed for iPhone and iPad devices. The digital 3D model was generated on cloud servers, after which it was possible to perform surface finishing (Adobe Photoshop) and a 3D model (3ds Max), in which the object was also rendered to transfer it to the Unity engine for presentation purposes.

SfM technology is used in extremely different fields, such as forestry [28] or medicine [29,30]. The SfM method was used to study three contrasting terrain forms at different scales, including: (i) exposed rocky coastal cliffs, (ii) a disturbed complex of moraine dams, and (iii) a bedrock ridge carved in the glacier [31]. Photogrammetric three-dimensional models of objects were made in Colombia, which served as a library of props in landscape visualisation [32]. The mapping technique was used in georeferencing of archaeological aerial photographs [33], and it was also used to study large monuments of the northern river terrace of the Uyuk valley (Russia) and their state of preservation was assessed on the basis of high-resolution optical satellite data [34]. The use of scanning and mapping techniques in the field of geography was used in the study of the water-level fluctuation zone in the Three Gorges Reservoir area in China. An innovative approach to photogrammetry using ships as a device carrier, which was also used for 3D acquisition of landslides, was proposed [35].

Photogrammetry techniques can also contribute to the restoration works of buildings. Study [36] discusses the activities related to the main characteristic elements of the Venetian basilica, which were documented using very high resolution orthophotos. The SfM method was used to integrate the data obtained from the Multi View Stereo and the images generated by a camera placed on an unmanned aerial system (UAS). Tests of generating 3D models based on ground and aerial data acquisition were carried out in the conservation of the Cistercian Staffarda Abbey in Piedmont, a cultural heritage site in Italy. Airborne data acquisition was based on a photogrammetric Remotely Piloted Aircraft System (RPAS) flight, while ground-based research was carried out by laser scanning. Both methods allowed to extract and process various point clouds and generate mesh 3D models of different scales (different resolution, detail content and precision) [37]. In turn, the obtained 3D point cloud was used to study the brick bridge over the Sele River (southern Italy), characterised by unique architecture of historical value, dating from the mid-nineteenth century [38].

Description of available free software for SfM technology and comparative studies of the use of four programs (COLMAP, OpenMVG, Theia and VisualSFM) to create 3D models of several different objects of sizes from 1 to several meters is presented in [2].

In the light of the works discussed, it appears that the application of the SfM method can be found in various fields: cultural heritage, museology, archeology, geography, as well as medicine and even forestry. The cited works also present other technologies of 3D data acquisition of digitised objects. They most often present case studies, and methodological issues are discussed marginally. There is a lack of research that would concern comparative analyses of 3D models obtained by various methods. Only one work was found [39], in which the authors compared the 3D models obtained using four different methods: structured light scanning, triangulation laser scanning, photometric stereo and close-range photogrammetry. This article attempts to answer questions about the efficiency and effectiveness of data acquisition methods and for the creation of 3D models as well as various applications in research and dissemination. Considering the repeatability, time and resource implications, qualitative and quantitative potential, and ease of use, studies of the strengths and weaknesses of these four methods are presented.

In the present work, an attempt is made to fill in the existing gap and present a comparison of the 3D SLS and SfM methods on the examples of digitisation of the same objects representing artefacts of material cultural heritage.

3. Method and Materials

3.1. Structured-Light 3D Scanning

Structured light scanners are active devices because they emit light beams from their own source. Scanners of this type project an appropriate pattern of light (a series of fringes with different and, in addition, time-varying widths, a dot matrix or other shape) onto the digitised object. Thanks to this, it is possible to accurately capture the geometric shape of the analysed surface. A light beam with a defined pattern is produced by built-in LCD projectors or other sources. In existing structures, white light is most often used, but also blue or green. These devices also have two or more cameras, as well as detectors that analyse and calculate the distance of surface points in the field of view. Thus the transferred object to the digital world has real dimensions. It should be emphasised, however, that scanners of this type are very sensitive to lighting conditions. All shadows on the real object are mapped onto the generated digital 3D model. For this reason, using them outside can be quite problematic. These scanners typically do not (or even not at all) capture data from black, shiny or transparent surfaces. In such situations the solution used is a matted surface by covering them with a special spray with powder before scanning, which is not allowed when digitising historic artefacts due to their protection. Scanners of this design are characterised by high speed of data acquisition and accuracy of measurements up to several micrometers. These scanners work at short distances (usually 15–40 cm) and, importantly, some structures (e.g., by Artec) can be hand-held, which significantly increases the efficiency of work when using them. There are also known solutions that always have to be mounted on tripods (e.g., RangeVision Neo devices).

Currently, there are many 3D SLS design solutions on the market. Spider and Eva model scanners (Artec, Luksemburg, Luksemburg) are frequently used devices in museum conditions. The fact that these models belong to one manufacturer means that they have common software and the same configuration process. Their main advantage is the comfort of use, because they are light and small, and thus allow a full range of motion during scanning. The advantages also include precise texture and colour reproduction as well as high resolution of the obtained point clouds. This allows for precise examination of objects of various sizes and varied surfaces. The Artec Eva model is better suited for scanning simple geometries, while the Artec Spider model for objects with a complex structure and intricate patterns. It is important that they do not require the use of positioning markers while working, which makes them ideal for scanning museum artefacts. There is also no need to calibrate them before each scanning process, which significantly simplifies and speeds up the work. The specifications of both scanners are presented in Table 1.

These scanners are an effective solution for safe and non-contact digitisation of museum collections. The light with which the measurements are made is neutral for them, so they can also be used to scan very delicate, old and valuable objects. Of course, there are other solutions for devices of this class. Study [40] presents data of the devices compared and parameters of the 3D models obtained.

Table 1. Specification of Artec Eva and Spider scanners *.

Parameters	Artec Eva Scanner	Artec Spider Scanner
Texture capture capability	yes	yes
Scanning technology	structured white light	flashing light, blue light (not laser)
Resolution 3D, up to	0.2 mm	0.1 mm
Point precision 3D, up to	0.1 mm	0.05 mm
3D precision at a distance, up to	0.03%/100 cm	0.03%/100 cm
Linear field of view, height × width	214 mm × 148 mm to 536 mm × 371 mm	90 mm × 70 mm to 180 mm × 140 mm
Working distance	0.4–1 m	0.17–0.35 m
Weight	0.85 kg	0.85 kg
Minimum requirements of the computer cooperating with the scanner	I5 (recommended I7), 8–12 GB RAM, NVIDIA GeForce 400 series	I5 (recommended I7), 8–12 GB RAM, NVIDIA GeForce 400 series

* The data come from the manufacturer, but the actual values do not differ from those given, which has been confirmed by our many years of use of this equipment.

3.2. Structure from Motion

The SfM method, belonging to the passive data acquisition technology, allows to create digital 3D models based on many two-dimensional images (photos). To generate a three-dimensional model, one needs a series of images of an area or object with a large overlap (about 80%), taken at different angles and at different distances, Figure 1 [41]. An average quality camera is sufficient to take pictures, and now even a camera built into a smartphone, making this method a low-cost solution. The smartphone is considered a device of common use, because its price in relation to the many different possibilities it offers to each owner is relatively low.

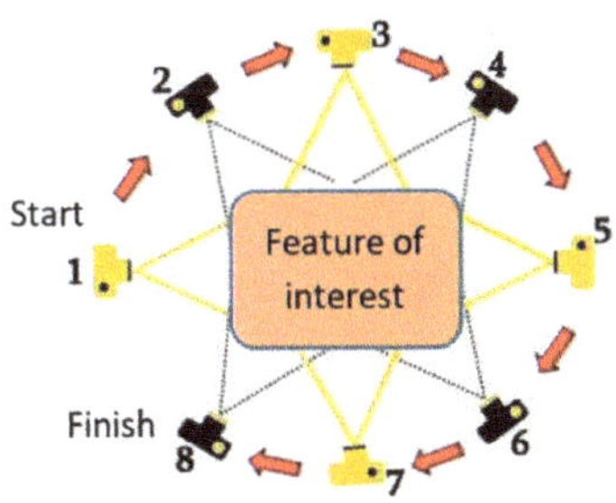

Figure 1. Taking a series of photos in the Structure from Motion method.

In the SfM method, a series of photos is a set of data that is imported into an appropriate computer program dedicated to this method. This software allows to generate 3D models and textures. Today, there are many programs available, both free and open-source, such as Meshroom [42,43], Colmap [44], VisualSFM [45], and commercial: Agisoft PhotoScan, Capturing Reality RealityCapture [2].

The general process of incremental SfM operations is shown in Figure 2 [44]. It can be broken down into two main modules: correspondence search and incremental reconstruction. The first module aims to search for complete and diffuse matches specific to two images. It has three sub-steps: feature extraction, feature matching, and geometric verification. The second module aims to estimate the alignment of images and three-dimensional scene points. It is possible thanks to the following steps: initialisation (selection of the best pair of images), image registration (for orientation), triangulation (calculation of 3D points), adjustment of the beam (local and global improvement) and removal of outliers.

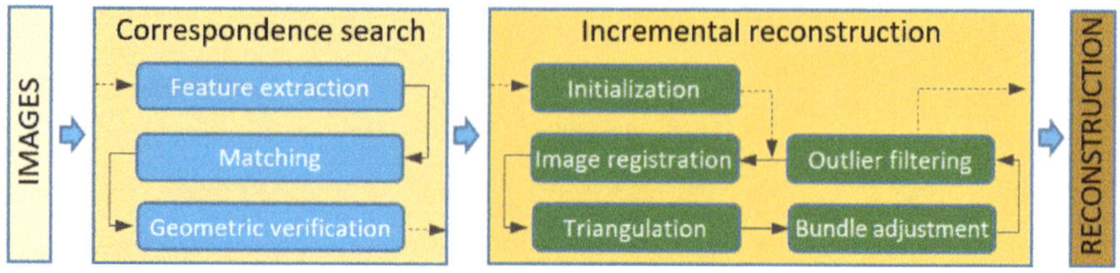

Figure 2. The process of operation of the software for the SfM method.

3.3. Description of the Objects of the Experiment

The Afrasiyab museum has exhibits from the city's excavation site. A trading city on the Silk Road, Samarkand was known from the 5th century BC, and was destroyed in 1220 by the invasion of Genghis Khan [46]. Currently, excavation sites covering an area of nearly 200 hectares are located within the administrative area of Samarkand (Uzbekistan). Selected exhibits from this museum have been described and presented using VR exposure [47]. Two very different objects were selected for the comparative analysis of digital three-dimensional models. The first object is a small glazed jug from the turn of the 11th–12th centuries [48]. The shape of the jug is typical of this period and region of origin, it is only 14.1 cm high and 8.9 cm in diameter. A characteristic feature of the jug is a long narrow funnel and a handle placed on the opposite side, of which, in this example, only the upper part of Figure 3a has been preserved. The second object is a reconstructed pottery kiln, 276 cm high, 219 cm wide and 110 cm deep. This object stands against the wall of the exhibition hall and is only accessible from the front, Figure 3b. To make its interior visible, a longitudinal section of the entire facility was made, and for exhibition purposes, jugs of various types and sizes were hung in several rows on the furnace wall, and several bowls and plates were placed inside.

Figure 3. Objects from the Afrasiyab museum: (**a**) 11th/12th century glazed jug, (**b**) reconstruction of the 9th–12th century pottery kiln.

3.4. 3D SLS Scanning and Postprocessing of the Jug and the Oven

3D scanning was performed in situ. Two handheld 3D SLSs were used for scanning: Artec Spider for digitising the jug and Artec Eva for scanning the furnace, beside a laptop with a 4-core processor and 16 GB of operating memory, Figure 4. Thanks to this, trouble-free and failure-free operation of scanners and quick processing of downloaded data was ensured. The dedicated software for Artec Studio scanners, version 12 Professional, was responsible for all activities related to data acquisition. The process of scanning the furnace, due to its large size and the complicated shapes of the vessels there, was significantly more difficult.

Figure 4. Scanning objects with SLS: (**a**) jug—Artec Spider, (**b**) model of kiln—Artec Eva.

Artec Studio software version 15 Professional and a computer equipped with an Intel Core i7 (3.4 GHz) processor, 64 GB RAM and an NVIDIA GeForce GTX 1080 graphics card with 8 GB internal memory were used to process the data obtained in the process of scanning the jug and stove. The stages of data processing in the form of 3D SLS point clouds on the example of a jug are shown in Figure 5.

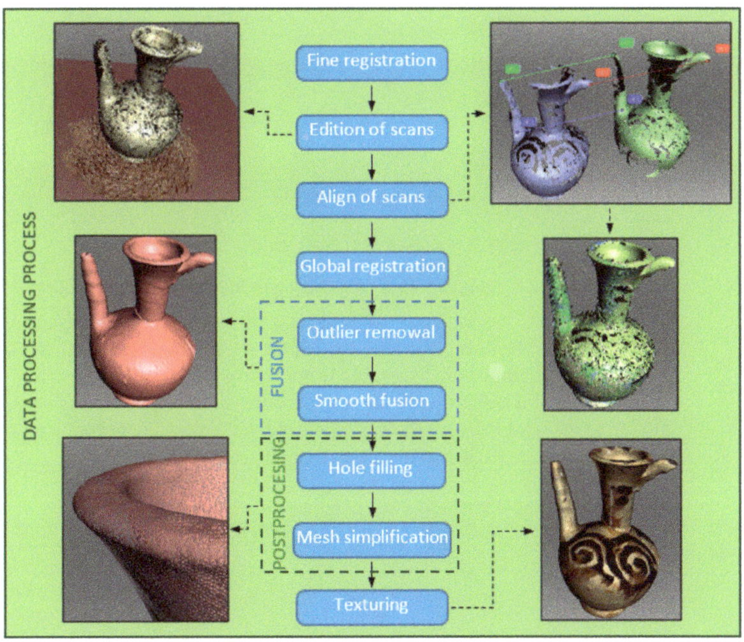

Figure 5. Stages of data processing obtained in the scanning process.

The final result of the post-processing is a mesh model of the jug and stove with the texture applied, which are exported to files in the .obj format. This format stores data about the spatial model, ie geometry, as well as information about its colour and texture (which is impossible when saving to the .stl format). The texture has been exported to a separate file and saved as a .jpg format. Models in .obj formations are described using polygons, Bezier curves and NURBS surfaces, which allows for seamless exchange and transfer between different applications for 3D graphics, minimising the risk of damage.

3.5. Structure from Motion Jug and Stove

The process of photographic documentation of the jug and the stove was carried out in situ by partners from Uzbekistan using a non-metric compact camera (D5300 digital SLR camera, Nikon, Tokio, Japan, equipped with a Nikkor lens with a focal length of 18–140). When taking pictures, the camera was moved in such a way that the following pictures cover more than 60% of the previous picture. Pictures were taken holding the camera in hand (without a tripod) at different distances and angles. They were saved in the .jpg format and had dimensions of 4496 × 4000 pixels and a resolution of 300 dpi (horizontally and vertically). Pictures of the jug were taken with the following camera parameters: exposure time 1/60–100 s, focal length 48–56 mm, maximum aperture 4.4–4.5. 49 photos were taken with a total size of 176 MB. In turn, 31 photos of the furnace (total size 112 MB) were taken with similar settings. The smaller number of photos was due to the fact that the stove was adjacent to the wall and was only visible from the front (Figure 3b).

The data processing (2D photographic images of the jug and the oven) to create three-dimensional mesh models and textures was carried out using the free Meshroom [42,43] v2020.1.1 version available under open-source license. The works were carried out on the same computer equipment as for the 3D SLS method. The steps of the data processing on the example of a jug are shown in Figure 6.

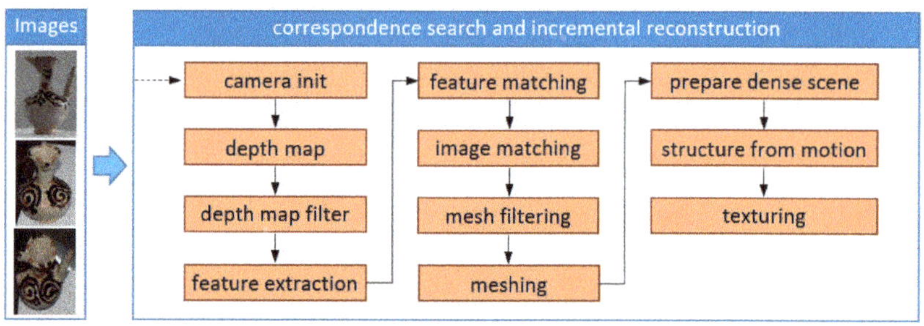

Figure 6. Stages of data processing in the SfM method.

3.6. Comparison of Data Acquisition Methods and Creating 3D Models

Acquiring data about a three-dimensional object using methods with completely different operating principles means that the user/museologist who would like to use them in practice must properly understand what is available and how it can be processed. Table 2 presents the main features and properties describing the process of data acquisition and processing to obtain digital 3D models of objects using Structured-light 3D Scanning (3D SLS) and Structure from Motion (SfM).

Table 2. Comparison of procedures for generating 3D models by the 3D SLS and SfM methods.

No	Features	3D SLS	SfM	Comments
1	Device	3D Scanner	Movie-camera/ camera/smartphone	
2	Shape data acquisition	3D scanning	Shooting	

Table 2. *Cont.*

No	Features	3D SLS	SfM	Comments
3	Texture data acquisition	RGB information *	2D photos	* selective uptake
4	Digitisation of transparent objects	No	Partially	
5	Data form	Point cloud	2D photos	
6	Data size	Several GB	Several to several hundred MB *	* Camera dependent
7	Software	Specialised, device-dedicated	Device-dedicated/Open source	
8	Computer hardware	High computing power	Low computing power	
9	3D model building method	Point cloud triangulation	Triangulation using image analysis	
10	3D model generation	Manual, assisted	Automated	
11	Costs	High/Very High	Low	
12	Availability	Low	Widespread	
13	Time consumption	Low/Medium	Low/Medium	Object-size dependent
14	Acquisition of dimensions	Yes	No/Yes *	* Necessity to place tags
15	Model quality	Very good	Small/medium	
16	Making it photorealistic	Texture mapping	Texture mapping	
17	Possibility to export to standard formats	Yes	Yes	
18	Possibility to modify models	Yes	Yes	
19	Competence in using the equipment	Specialist/High	Common/Medium	
20	IT competences	Specialist/High	Common/Medium	

4. Results

The obtained results are divided into digitised objects (jug, oven), showing the visual effects of the generated 3D models, after using the 3D SLS and SfM methods, before and after their processing, and the tables present detailed parameter values describing mesh models and textures.

4.1. Digital 3D Models of the Jug

4.1.1. 3D SLS and Postprocessing

When scanning the jug (Artec Spider), three scans were made at different positions of the object on a manually rotated table covered with a special texture ensuring proper positioning of subsequent points of the surface taken from the object (Figure 4a). The resulting files containing all point clouds and object texture information were 1.12 GB in size. Information on individual scans is presented in Table 3.

Table 3. Data on scans for an object like a jug.

Scan No.	Surface	Polygons	Vertices	Frame	Texture
1	1060	23,169,773	13,514,745	1060	59
2	280	9,848,052	5,612,084	280	28
3	496	12,634,644	7,145,096	496	115
Total	1836	45,652,469	26,271,925	1836	202

After carrying out all stages of the postprocessing (Figure 5), a mesh model with 214,870 vertices and 429,736 faces was obtained. The size of the .obj file was 30.1 MB, while the texture for surface mapping with dimensions of 4096 × 4096 pixels and the horizontal and vertical resolution of 96 dpi saved in the .jpg format was 7.1 MB. The results of the SLS scanning of the jug are shown in Figure 7a,b.

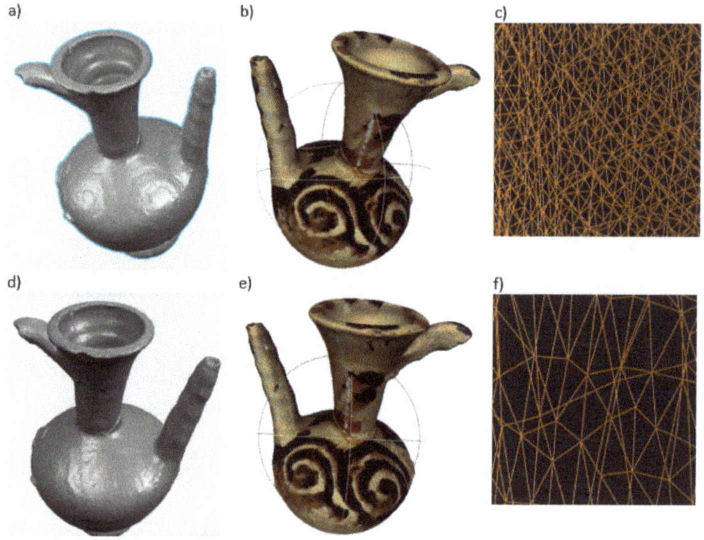

Figure 7. The appearance of a digital 3D model of a jug with 3D SLS: (**a**,**b**) primary model—without and with texture, (**d**,**e**) model after additional processing—without and with texture, (**c**,**f**) mesh view before and after optimization.

Due to the large number of vertices and faces in the model grid and the large size of the .obj file (Table 4), the mesh was reduced using the free Blender program. For this purpose, the Decimate tool was used, which allows to change the coordinates of the vertices, and thus the walls, in relation to the original model. The Decimate tool permits one to modify the mesh of the model using one of the methods: Collapse, Un-Subdivide and Planar. Based on the experience gained earlier, the authors used the Collapse method. The actions performed led to a tenfold reduction of the model mesh, resulting in 21,488 vertices and 42,972 faces (Figure 7c,f). This resulted in the size of the .obj file being reduced to 4.31 MB. The texture has also been changed from 4096 × 4096 px to 2048 × 2048 px. The file size after .jpg optimisation was 953 kB. Despite the use of such large simplifications, they did not significantly deteriorate the quality of the model and its texture. It is true that there is no spiral in Figure 7d, which is visible in the base drawing (Figure 7a), but after applying the texture, the differences between the drawings are generally not visible (Figure 7b,e), and it is the 3D models covered with textures which are presented in virtual museums.

Table 4. Parameters of the digital 3D model and the texture of the jug.

	Model Mesh			Texture		
	Vertices	Faces	.obj File Size [MB]	.jpg File Size [MB]	Size [Pixels]	Resolution [dpi]
	3D SLS method					
Original model	214,870	429,736	30.1	7.1 MB	4096 × 4096	96 × 96
Processed model	21,488	42,972	4.31	953 kB	2048 × 2048	96 × 96
	SfM method					
Without processing photos	119,694	238,712	16.1	5.31 MB	4096 × 4096	96 × 96
After processing photos	102,369	204,486	13.0	3.82 MB	4096 × 4096	96 × 96
After smoothing the model's surface	102,369	204,486	21.4	3.82 MB	4096 × 4096	96 × 96

4.1.2. SfM and Postprocessing

The procedure shown in Figure 6 for creating a 3D model of the jug by the SfM method was used in two variants.

Option 1—use of original photos, without digital processing, see Figure 8.

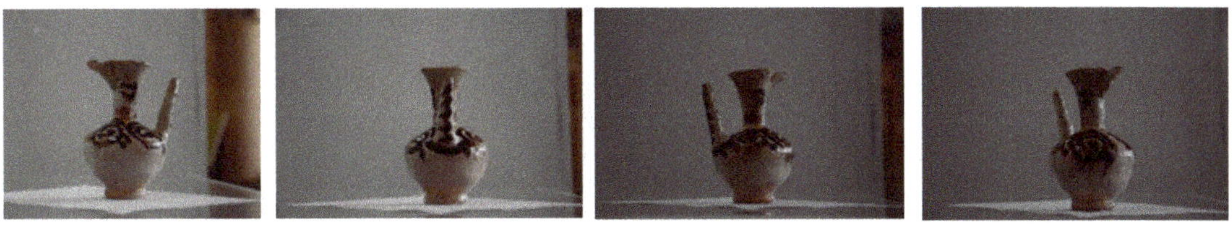

Figure 8. Sample photos of the jug without processing received from partners from Uzbekistan.

After 49 photos were uploaded to Meshroom and verified by the software, only 28 photos were qualified for further analysis. After performing all steps of the procedure, the mesh model with the applied texture was obtained. The model was exported to a .obj file and the texture to a .jpg file. The appearance of the digital 3D model obtained from the original photos (untreated) is shown in Figure 9. It is clearly visible that on one side of the model the geometric surfaces and the texture were very well recreated (Figure 9a,c). This can be seen on the surface of the spout, neck and body of the jug. The edges of the spout are even and not jagged. On the other side of the model, however, there are many surface discontinuities, the mesh is heavily deformed (no smoothing, there are many outliers from the background of the photos, as shown by the arrows in Figure 9d). Outliers have been incorporated into the mesh causing additional distortions of the geometric surfaces of the model.

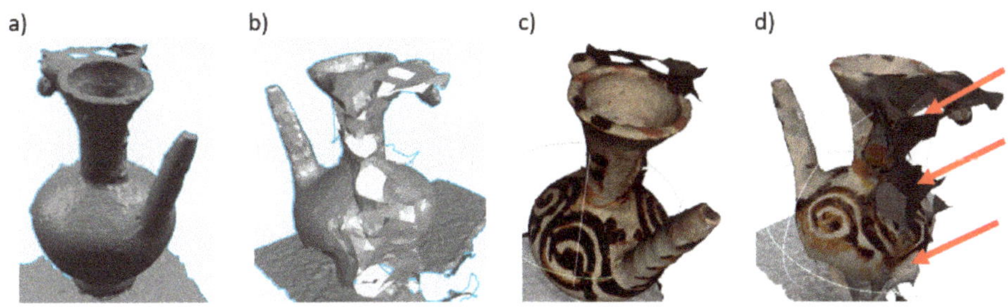

Figure 9. The appearance of a digital model of a jug with SfM: (**a**,**b**) without texture—side one and two, (**c**,**d**) with texture—side one and two.

Option 2—using photos after processing, Figure 10.

Figure 10. Sample photos after post-processing.

The processing of the obtained photos consisted of cropping them and correcting the contrast and brightness in order to obtain objects with similar colours. Framing was carried out in such a way that the jug was located.

In this case, after loading 49 photos into Meshroom, only eight of them were rejected. Thus, the process of creating a 3D model was carried out using 41 photos. The obtained mesh model is shown in Figure 11. It is not difficult to notice that the obtained three-dimensional mesh model of the jug was significantly improved, all sides of the object were correctly recreated. Only some surface irregularities appeared on the spout and the walls of the neck and funnel (as shown by arrows in Figure 11b,d).

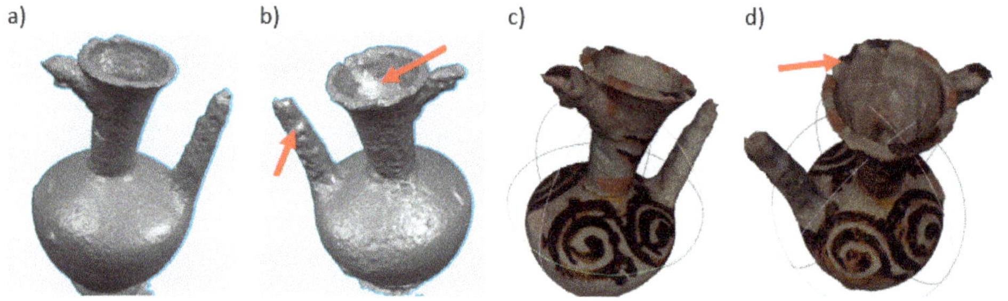

Figure 11. The appearance of the digital model of the jug obtained by the SfM method after photo processing: (**a**,**b**) without texture—page one and two, (**c**,**d**) with texture—page one and two.

The improvement of the roughness of the jug surface was achieved by a smoothing process (Smoth modifier) performed with the use of the free Blender software, Figure 12.

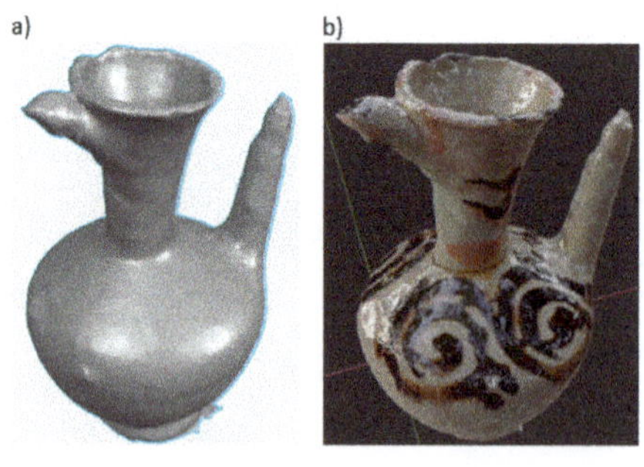

Figure 12. The appearance of the digital 3D model of the jug obtained by the SfM method after smoothing the surface: (**a**) without texture, (**b**) with texture.

4.1.3. Comparison of SLS and SfM Technologies

Collective parameters of digital 3D models (geometry and texture) of a pitcher made in SLS and SfM technology are presented in Table 4.

The analysis of the results in Table 4 and the visualisation from Figure 7 shows that in the case of the 3D model obtained from SLS after the 10-fold mesh reduction and 2-fold reduction of the texture dimension the visual quality of the model was maintained and the total file size decreased by more than eight times. This makes it possible to present such a model on websites. In the case of digital objects from SfM technology, the situation is different. The resulting primary model (with photo processing) has the total size (mesh model and texture) more than twice smaller than the SLS model. The surface smoothing process (with the mesh count unchanged) improved the visual quality of the model, but there was a significant increase in the size of its mesh model (by over 60%). Despite the still inferior visual quality of this model, its total size is approximately five times larger than that of the SLS model. Thus it seems that its suitability for presentation on the web becomes more problematic.

4.2. Digital Oven 3D Models

4.2.1. 3D SLS and Postprocessing

After digitising with the Artec Eva scanner (Figure 4b) 28 scans were obtained. Their large number was due to the size of the furnace (approximately 2.8 × 2.2 m). The size of the files with the scans obtained was 7.2 GB. In total, 13,967 frames were received, over 416 million polygons, over 221 million vertices and 1029 datapoints about texture. After carrying out the data processing process in accordance with the diagram shown in Figure 5, a 3D mesh model with applied texture was obtained. As in the case of the jug model, it was exported to a file in .obj format, and the texture was saved in the file in .jpg format. A view of the furnace and its components is shown in Figures 13a and 14.

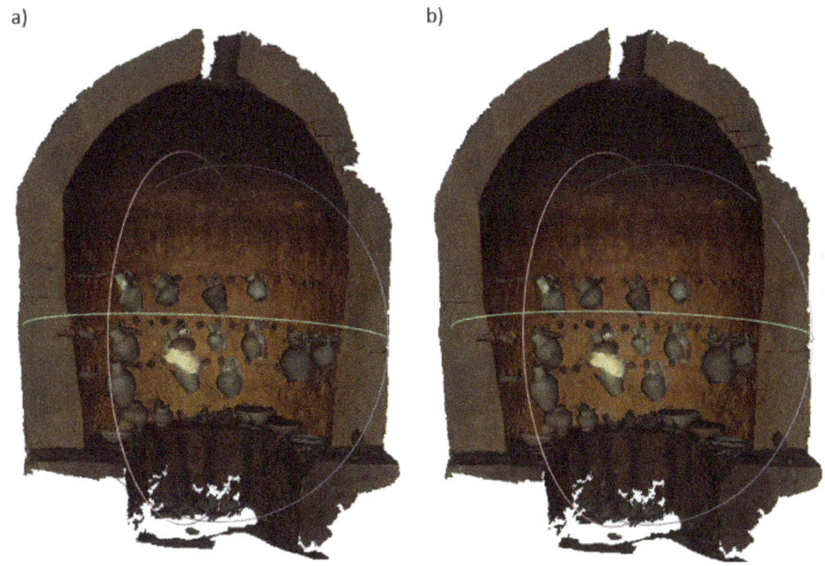

Figure 13. Three-dimensional model of the furnace (**a**) original model, (**b**) after optimalisation.

Figure 14. A 3D model with the texture of objects inside the furnace.

Most of the elements that were placed in the centre of the furnace were very well recreated. The 3D models are of very good quality. However, there is an area (shown by the ellipse in Figure 14) where the three-dimensional models have not been properly formed. The presented case results from the fact that in the scanning process the identification of consecutively collected points in relation to the previously recorded points was lost. Such a situation can be identified in the process of point cloud processing. However, for the purposes of this article, this has not been done to show the visual consequences of this situation.

The optimisation of the 3D model of the furnace consisted in simplifying the mesh and texture: the mesh was reduced ten times and the dimensions of the texture used were reduced four times. The data obtained were recorded in Table 5, the realistic model is shown in Figure 13b. After applying mesh optimisation, no significant qualitative differences were noticed compared to the original model.

Table 5. Parameters of the digital 3D model and the texture of the furnace.

	Model Mesh			Texture		
	Vertices	**Faces**	**.obj File Size [MB]**	**.jpg File Size [MB]**	**Size [Pixels]**	**Resolution [dpi]**
		3D SLS method				
Original model	2,110,146	4,199,231	311	30.4	8192 × 8192	96 × 96
After optimisation	218,819	419,922	46.7	1.12	2048 × 2048	96 × 96
		SfM method				
Original model	717,745	1,434,925	98.9	12.5	4096 × 4096	96 × 96
After smoothing the model's surface	717,745	1,434,925	160	12.5	4096 × 4096	96 × 96

4.2.2. SfM and Postprocessing

The procedure shown in Figure 6 was also used to create a three-dimensional model of the furnace by the SfM method. Out of 31 original photos loaded (without processing), Meshroom, after verification, used all photos for further analysis. The obtained three-dimensional model of the furnace is shown in Figure 15a and its elements in Figure 16a.

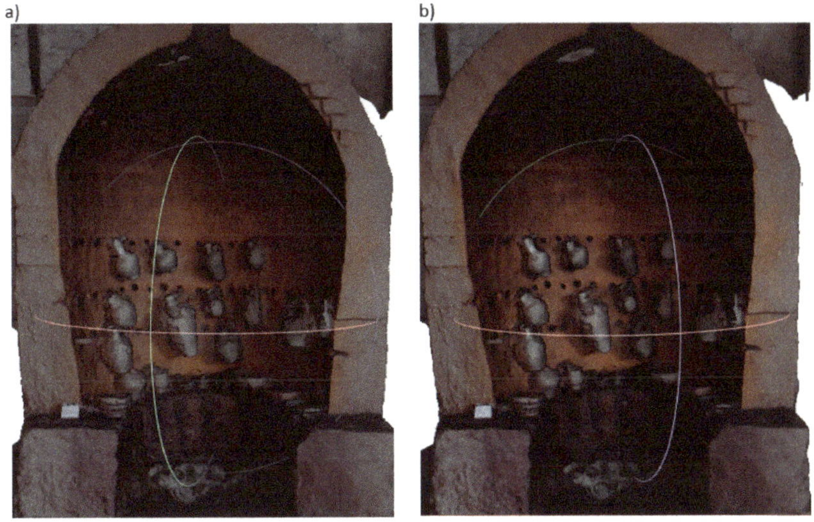

Figure 15. Three-dimensional model of the furnace obtained by the SfM method: (**a**) original version, (**b**) after smoothing the surface.

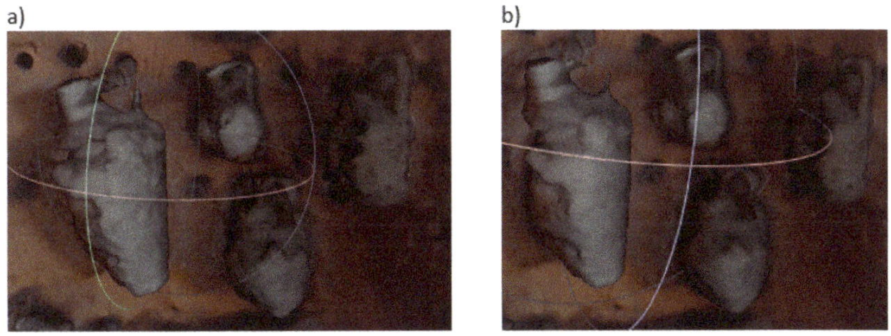

Figure 16. 3D model with the texture of the dishes in the furnace: (**a**) original version, (**b**) after smoothing the surface.

While the quality of the obtained model of the furnace itself is quite good, the models of vessels in it are not satisfactory. They are definitely different from the models obtained by the 3D SLS method. For this, the 3D model underwent a surface smoothing process in Blender. After carrying out this operation, the unevenness on the surfaces of the furnace itself (Figure 15b), as well as on the surfaces of the elements inside it, decreased (Figure 16b).

4.2.3. Comparison of SLS and SfM Technologies

The collective parameters of digital 3D models of the furnace made in SLS and SfM technology are presented in Table 5.

The results included in Table 5 show that the optimisation of the 3D model obtained from the SLS method, consisting in a 10-fold reduction of the mesh size and 4-fold reduction of the texture dimension, with a virtually invisible visual change of the object, resulted in a reduction of the total file size over 6.5 times. With regard to the digital model from the SfM method, the improvement in object visibility after smoothing resulted in an almost 2-fold increase in the .obj file size. It should be added that the size of this file is almost 3.5 times larger than the SLS file, and the appearance of this model is of lower quality anyway.

Tables 6 and 7 present the time-consuming nature of data acquisition and processing when creating digital 3D models of a jug and a stove using the SfM and 3D SLS methods. Table 8 lists the costs of devices and software used in the SfM and 3D SLS methods.

Table 6. Time-consuming production of the 3D model of the jug using the SfM and 3D SLS methods.

<table>
<tr><th></th><th>SfM/Meshroom Program</th><th>min</th><th>3D SLS/Artec Studio Professional 15 Program</th><th>min</th></tr>
<tr><td>Data Acquisition</td><td></td><td>12</td><td></td><td>5</td></tr>
<tr><td rowspan="5">Data processing</td><td>Making a 3D model from raw photos—an automated process</td><td>20</td><td rowspan="3">Converting a point cloud into a 3D mesh model—manual process</td><td rowspan="3">70</td></tr>
<tr><td>Photo processing</td><td>90</td></tr>
<tr><td>Making a 3D model from processed photos</td><td>20</td></tr>
<tr><td>Smoothing the model in Blender</td><td>8</td><td>Optimising the 3D model mesh in Blender</td><td>10</td></tr>
<tr><td>Total</td><td>150</td><td>Total</td><td>85</td></tr>
</table>

Table 7. Time-consuming production of 3D models of the furnace using the SfM and 3D SLS methods.

<table>
<tr><th></th><th>SfM/Meshroom Program</th><th>min</th><th>3D SLS/Artec Studio Professional 15 Program</th><th>min</th></tr>
<tr><td>Data Acquisition</td><td></td><td>3</td><td></td><td>55</td></tr>
<tr><td rowspan="3">Data processing</td><td>Making a 3D model from raw photos—an automated process</td><td>20</td><td>Converting a point cloud into a 3D mesh model—manual process</td><td>480</td></tr>
<tr><td>Smoothing the model in Blender</td><td>15</td><td>Optimising the 3D model mesh in Blender</td><td>15</td></tr>
<tr><td>Total</td><td>38</td><td>Total</td><td>550</td></tr>
</table>

Table 8. List of hardware and software costs.

<table>
<tr><th></th><th>SfM *</th><th></th><th>3D SLS **</th><th></th></tr>
<tr><td>Cost</td><td></td><td>€</td><td></td><td>€</td></tr>
<tr><td rowspan="2">Device</td><td rowspan="2">Nikon camera with Nikkor</td><td rowspan="2">970</td><td>Artec Spider Scanner</td><td>19,700</td></tr>
<tr><td>Artec Eva Scanner</td><td>13,700</td></tr>
<tr><td rowspan="2">Software</td><td>Meshroom</td><td>free</td><td>Artec Studio 15 Professional</td><td>2000</td></tr>
<tr><td>Blender</td><td>free</td><td>Blender</td><td>free</td></tr>
</table>

* excluding computer hardware for data processing. ** excluding computer hardware for scanning and data processing.

The analysis of the results in Table 6 concerning the generation of a 3D model of a small jug shows that the time taken by the SfM method is almost twice as long as that of the 3D SLS methods. If more photos were taken, the time would increase to about 3 times. Knowing that raw photos need to be pre-processed, time can be saved by generating a 3D model from these photos and cropping them can follow straight away. In the case of a large object—a furnace (Table 7), it can be seen that the SfM method is about 15 times less time-consuming than the 3D SLS methods. In the method, scanning a large, complex surface takes a long time, as opposed to taking photos. When photographing the furnace, there was no need to create a shooting station, and moreover, the furnace was positioned against the wall, so only one accessible side was photographed. All this meant that the shooting time was 4 times shorter than that of a small jug. The program generating a 3D model from photos requires approximately the same time to generate the model, regardless of whether the object in the photos is large or small. Initial processing of scans from the 3D SLS method, consisting in separating point clouds from each other, when information about reference points was lost in the scanning process, and their recording as separate post-drafts must be performed manually. This operation is very time-consuming, but allows to keep all the data obtained during the scan.

The results from Table 8 show that the costs of professional equipment for the 3D SLS method compared to the purchase of a camera are huge (over 35 times) and many museum institutions will not be able to afford them. However, it should be remembered that acquiring a point cloud of the scanned object ensures the archiving of raw data, which constitute the so-called perpetual archiving of a museum artefact.

5. Conclusions and Further Work

The article presents a comparison of the results of safe and non-contact digitisation of objects from museum exhibitions using two different methods: the 3d Structural-Light Scanning (3D SLS) and the Structure from Motion (SfM) methods. It has been shown that it is possible to select programs and tools, as well as devices enabling the creation of low-cost 3D mesh models using the SfM method, using free software (Meshroom, Blender) and commonly owned devices and computer equipment (camera, laptop).

Using modern information technologies, it has been shown that it is possible to carry out works when employees of partner units are located in different geographic locations in Lublin (Poland) and Samarkand (Uzbekistan). Each team can focus on activities related to professional preparation, sharing their knowledge at the same time, in order to gain new competences in the future in the field of creating digital copies of real models or tangible cultural heritage. By combining the competences of both teams, a synergy effect occurs, which makes it easier to solve emerging problems.

In the light of the research consisting in the creation of digital 3D models of museum objects using the 3D SLS and SfM methods and a comparative analysis of the obtained results, it became possible to formulate the following conclusions:

(1) Scanning with the 3D SLS method with the use of Artec Spider and Artec Eva scanners is perfect for mapping museum objects due to the lack of contact with them (the emitted light is neutral for the scanned objects, thanks to which it is possible to scan very delicate, old and valuable objects). It enables effective in situ data acquisition, which significantly simplifies the digitisation process and the necessary formalities to carry it out.

(2) Performing the proper postprocessing of the data obtained in the 3D SLS scanning process allows to obtain a faithful 3D mesh model with the applied texture, which can be exported to a file in .obj format with a size many times smaller than the original data obtained.

(3) Optimisation of the 3D model mesh using the 3D SLS method and reducing it by up to 10 times while reducing the texture dimension by a factor of 2 does not result in the loss of good visual quality of these objects. This remark applies to very small objects (a 14 cm high jug and a 270 cm high oven). It should be added that the total size of model and texture files is reduced by about 6 to 8 times. As a result, these models are well suited for placement on museum websites.

(4) The low-cost SfM method has an unquestionable advantage (over 35 times), which can be an attractive alternative when creating digital 3D models of museum objects. The results of the comparative studies, however, show some of its shortcomings, which means that the obtained three-dimensional digital models are usually of lower visual quality. In addition, due to the improvement of their quality (e.g., surface smoothing), their size increases significantly and is about 3–5 times larger than the size of model files and texture files from the 3D SLS method after optimisation. This may mean that inserting such large model files on museum websites may discourage potential users due to the long download times.

(5) The use of the SfM method to create digital 3D models of objects of medium or small size (e.g., a jug with a height of 14 cm) gives much better results when the process of cropping the photo is carried out so that the object is in the centre of the photo. The amount of noise generated in the background of the photo is reduced.

(6) Although the 3D models generated by the SfM method were obtained from a small number of photos, about 30–50 (photos taken by partners from Uzbekistan), the quality of these models turned out to be quite good. This is especially true of a small jug—an object in its main axisymmetric shape. So attention should be paid to getting much more photos of the subject (from 80 to 140).
(7) The time of the SfM method when applied to small objects can be reduced (more than 4 times) with careful photographing, paying attention that the object is well-framed, without unnecessary background of the surroundings.
(8) The classic SfM method does not provide for placing markers next to the photographed object, so the obtained digital 3D models do not have information about their dimensions. Thus, these models are not fully suitable for the professional archiving of historical objects, but may be useful for popularising and making available tangible cultural heritage.

The authors see the need to continue the work in terms of:

(1) Conducting comparative studies of creating digital 3D models of museum objects using the SfM method with the use of other non-commercial and commercial programs.
(2) Development of methods and algorithms for comparing the obtained 3D models, which make it possible to calculate the differences between the shape of digital artefacts obtained from both methods and to change the shape of the surface of smoothed and optimised objects in relation to the basic model obtained by the 3D SLS method.
(3) Searching for better methods of post-processing 3D models from SfM to improve their visual quality while reducing their size.
(4) Developing a list of good practices for the preparation of a collection of photographs of museum objects for the purposes of the SfM method for museum worker.

Author Contributions: Conceptualization, J.M. and M.B.; methodology, J.M., M.B., M.P.-Ł. and A.S.; software, A.S., M.B.; validation, M.B. and A.S.; formal analysis, J.M.; investigation, J.M.; resources, J.M., M.B. and M.P.-Ł.; data curation, J.M., M.B. and M.P.-Ł.; writing—original draft preparation, J.M. and M.B.; writing—review and editing, J.M., M.B. and M.P.-Ł.; visualization, M.B.; supervision, M.B. and M.P.-Ł.; project administration, J.M. and M.B.; funding acquisition, J.M. All authors have read and agreed to the published version of the manuscript.

Funding: This article has been supported by the Polish National Agency for Academic Exchange under Grant No. PPI/APM/2019/1/00004 titled "3D DIGITAL SILK ROAD".

Institutional Review Board Statement: Not applicable.

Informed Consent Statement: Not applicable.

Data Availability Statement: Not applicable.

Acknowledgments: The authors would like to thank Samariddin Mustafokulov—Director of the Afrasiyab Museum in Samarkand for making it possible for us to 3D scan museum resources in 2017 and Daniyarov Fayzullo—the new Director of the Afrasiyab Museum, as well as the employees of the Museum for providing a collection of photos of objects from the exhibition.

Conflicts of Interest: The authors declare no conflict of interest.

References

1. Bendicho, V.M.L.-M.; Marchante-Ortega, Á.; Vincent, M.; Martín-Buitrago, Á.J.C.; Pintado, J.O. Uso combinado de la fotografía digital nocturna y de la fotogrametría en los procesos de documentación de petroglifos: El caso de Alcázar de San Juan (Ciudad Real, España). *Virtual Archaeol. Rev.* **2017**, *8*, 64–74. [CrossRef]
2. Bianco, S.; Ciocca, G.; Marelli, D. Evaluating the Performance of Structure from Motion Pipelines. *J. Imaging* **2018**, *4*, 98. [CrossRef]
3. COVID-19 and Tourism: Assessing the Economic Consequences, United Nations Conference on Trade and Development. 2020. Available online: https://unctad.org/system/files/official-document/ditcinf2020d3_en.pdf (accessed on 1 May 2021).
4. Miłosz, M.; Montusiewicz, J.; Kęsik, J. 3d information technologies in cultural heritage preservation and popularization-a series 602 of seminars for museologists made by computer scientists. In *Proceedings of the EDULEARN 20 Proceedings*; IATED: Walencja, Spain, 2020; pp. 544–549. [CrossRef]

5. Polish National Agency for Academic Exchange under Grant No. PPI/APM/2019/1/00004 Titled "3D DIGITAL SILK ROAD". Available online: https://nawa.gov.pl/en (accessed on 5 June 2021).
6. Miłosz, M.; Kęsik, J.; Montusiewicz, J. 3D Scanning and Visualization of Large Monuments of Timurid Architecture in Cen-tral Asia—A Methodical Approach. *J. Comput. Cult. Herit.* **2020**, *14*, 1–31. [CrossRef]
7. Montusiewicz, J.; Barszcz, M.; Dziedzic, K. Photorealistic 3D Digital Reconstruction of a Clay Pitcher. *Adv. Sci. Technol. Res. J.* **2019**, *13*, 255–263. [CrossRef]
8. Horn, C.; Ling, J.; Bertilsson, U.; Potter, R. By All Means Necessary—2.5D and 3D Recording of Surfaces in the Study of Southern Scandinavian Rock Art. *Open Archaeol.* **2018**, *4*, 81–96. [CrossRef]
9. Balletti, C.; Ballarin, M. An Application of Integrated 3D Technologies for Replicas in Cultural Heritage. *ISPRS Int. J. Geo-Inf.* **2019**, *8*, 285. [CrossRef]
10. Montusiewicz, J.; Barszcz, M.; Dziedzic, K.; Nowicki, T. The Method of Decomposition of Architectural Objects for the prep-aration of 3D Virtual Models and Replication. *Adv. Sci. Technol. Res. J.* **2021**, *15*, 247–257. [CrossRef]
11. Wachowiak, M.J.; Karas, B.V. 3D Scanning and Replication for Museum and Cultural Heritage Applications. *J. Am. Inst. Conserv.* **2009**, *48*, 141–158. [CrossRef]
12. Younan, S.; Treadaway, C. Digital 3D models of heritage artefacts: Towards a digital dream space. *Digit. Appl. Archaeol. Cult. Herit.* **2015**, *2*, 240–247. [CrossRef]
13. Balletti, C.; Guerra, F.; Scocca, V.; Gottardi, C. 3D integrated methodologies for the documentation and the virtual recon-struction of an archaeological site. In Proceedings of the International Archives of the Photogrammetry, Remote Sensing and Spatial Information Sciences, 3D Virtual Reconstruction and Visualization of Complex Architectures, Avila, Spain, 25–27 February 2015; Volume XL-5/W4, pp. 215–222.
14. Douglass, M.; Lin, S.; Chodoronek, M. The Application of 3D Photogrammetry for In-Field Documentation of Archaeologi-cal Features. *Adv. Archaeol. Pract.* **2015**, *3*, 136–152. [CrossRef]
15. Howland, M.D.; Kuester, F.; Levy, T.E. Photogrammetry in the field: Documenting, recording, and presenting archaeology. *Mediter. Archaeol. Archaeom.* **2014**, *4*, 101–108.
16. Grechi, G.; Fiorucci, M.; Marmoni, G.; Martino, S. 3D Thermal Monitoring of Jointed Rock Masses through Infrared Thermography and Photogrammetry. *Remote Sens.* **2021**, *13*, 957. [CrossRef]
17. Kong, D.; Saroglou, C.; Wu, F.; Sha, P.; Li, B. Development and application of UAV-SfM photogrammetry for quantitative characterization of rock mass discontinuities. *Int. J. Rock Mech. Min. Sci.* **2021**, *141*, 1–19. [CrossRef]
18. Lambers, K.; Eisenbeiss, H.; Sauerbier, M.; Kupferschmidt, D.; Gaisecker, T.; Sotoodeh, S.; Hanusch, T. Combining photo-grammetry and laser scanning for the recording and modeling of the Late Intermediate period site of Pinchango Alto, Pal-pa, Peru. *J. Archaeol. Sci.* **2017**, *34*, 1702–1712. [CrossRef]
19. Bruno, F.; Bruno, S.; De Sensi, G.; Luchi, M.-L.; Mancuso, S.; Muzzupappa, M. From 3D reconstruction to virtual reality: A complete methodology for digital archaeological exhibition. *J. Cult. Herit.* **2010**, *11*, 42–49. [CrossRef]
20. Lerma, J.L.; Muir, C. Evaluating the 3D documentation of an early Christian upright stone with carvings from Scotland with multiples images. *J. Archaeol. Sci.* **2014**, *46*, 311–318. [CrossRef]
21. Wei, O.C.; Chin, C.S.; Majid, Z.; Setan, H. 3D documentation and preservation of historical monument using terrestrial laser scanning. *Geoinf. Sci. J.* **2010**, *10*, 73–90.
22. David, A.; Kaminski, J. 3D scanning and presentation of ethnographic collections—Potentials and challenges. *J. Mus. Ethnogr.* **2014**, *27*, 78–97.
23. Higueras, M.; Calero, A.I.; Collado-Montero, F.J. Digital 3D modeling using photogrammetry and 3D printing applied to the restoration of a Hispano-Roman architectural ornament. *Digit. Appl. Archaeol. Cult. Herit.* **2021**, *20*, 1–11.
24. Solter, A.; Gajski, D. Project "towards the virtual museum"—Exploring tools and methods for 3D digitalization and visual-ization. *Opvscvla Archaeol.* **2018**, *39–40*, 117–124.
25. Tucci, G.; Cini, D.; Nobile, A. Effective 3D digitization of archaeological artifacts for interactive virtual museum. *ISPRS Int. Arch. Photogramm. Remote Sens. Spat. Inf. Sci.* **2012**, *XXXVIII-5*, 413–420. [CrossRef]
26. Neamtu, C.; Comes, R. Methodology to create digital and virtual 3d artefacts in archaeology. *J. Anc. Hist. Archaeol.* **2016**, *3*, 65–74. [CrossRef]
27. Capacete-Caballero, X.; Caulfield-Sriklad, D.; McKay, F. Enhancing the display of the fashion artefact through digital mul-ti-media approaches. In Proceedings of the 1st International Conference on Digital Fashion, London, UK, 16–17 May 2013; pp. 336–345.
28. Iglhaut, J.; Cabo, C.; Puliti, S.; Piermattei, L.; O'Connor, J.; Rosette, J. Structure from Motion Photogrammetry in Forestry: A Review. *Curr. For. Rep.* **2019**, *5*, 155–168. [CrossRef]
29. Dappa, E.; Higashigaito, K.; Fornaro, J.; Leschka, S.; Wildermuth, S.; Alkadhi, H. Cinematic rendering—An alternative to volume rendering for 3D computed tomography imaging. *Insights Imaging* **2016**, *7*, 849–856. [CrossRef] [PubMed]
30. De Moraes Jorge, A.P.; Monteiro, E.R.; Hoogenboom, B.J.; Oliveira, A.; Quintela, M.V.P. Computer photo-grammetry as a postural assessment in Schwartz-Jampel syndrome: A case report. *J. Bodyw. Mov. Ther.* **2021**, *26*, 72–76. [CrossRef] [PubMed]
31. Westoby, M.J.; Brasington, J.; Glasser, N.F.; Hambrey, M.J.; Reynolds, J.M. "Structure-from-Motion" photogrammetry: A low-cost, effective tool for geoscience applications. *Geomorphology* **2012**, *179*, 300–314. [CrossRef]

32. Silva-Bolíva, J.; Cataño-Ospina, A.M.; Arenas-Becerra, L.Y. Photogrammetry for the reconstruction of realistic visual landscapes that serve for the creation of scenographies in audiovisual and multimedia products. *J. Phys. Conf. Ser.* **2019**, *1418*, 23–25. [CrossRef]
33. Verhoeven, G.; Doneus, M.; Briese, C.; Vermeulen, F. Mapping by matching: A computer vision-based approach to fast and accurate georeferencing of archaeological aerial photographs. *J. Archaeol. Sci.* **2012**, *39*, 2060–2070. [CrossRef]
34. Caspari, G. Mapping and Damage Assessment of "Royal" Burial Mounds in the Siberian Valley of the Kings. *Remote Sens.* **2020**, *12*, 773. [CrossRef]
35. Jin, D.; Li, J.; Gong, J.; Li, Y.; Zhao, Z.; Li, Y.; Li, D.; Yu, K.; Wang, S. Shipborne Mobile Photogrammetry for 3D Mapping and Landslide Detection of the Water-Level Fluctuation Zone in the Three Gorges Reservoir Area, China. *Remote Sens.* **2021**, *13*, 1007. [CrossRef]
36. Fregonese, L.; Adami, A. The 3D Model of St. Mark's Basilica in Venice. In *Digital Transformation of the Design, Construction and Management Processes of the Built Environment. Research for Development*; Daniotti, B., Gianinetto, M., Della Torre, S., Eds.; Springer: Cham, Switzerland, 2020; pp. 343–354.
37. Bastonero, P.; Donadio, E.; Chiabrando, F.; Spano, A.T. Fusion of 3D models derived from TLS and image-based techniques for CH enhanced documentation. *ISPRS Ann. Photogramm. Remote Sens. Spat. Inf. Sci.* **2014**, *II-5*, 73–80. [CrossRef]
38. Pepe, M.; Costantino, D. UAV Photogrammetry and 3D Modelling of Complex Architecture for Maintenance Purposes: The Case Study of the Masonry Bridge on the Sele River, Italy. *Period. Polytech. Civ. Eng.* **2020**, *65*, 191–203. [CrossRef]
39. Graham, C.A.; Akoglu, K.G.; Lassen, A.W.; Simon, S. Epic dimensions: A comparative analysis of 3d acquisition methods. *ISPRS Int. Arch. Photogramm. Remote Sens. Spat. Inf. Sci.* **2017**, *XLII-2/W5*, 287–293. [CrossRef]
40. Kersten, T.P.; Lindstaedt, M.; Starosta, D. Comparative geometrical accuracy investigations of hand-held 3D scanning systems—An update. *Int. Arch. Photogramm. Remote Sens. Spat. Inf. Sci.* **2018**, *42*. [CrossRef]
41. Doumit, J.; Kiselev, E. Structure from motion technology from micro scale objects cartography. *Earth Sci.* **2017**, 42–47.
42. Hellmuth, R.; Wehner, F.; Giannakidis, A. Datasets of captured images of three different devices for photogrammetry cal-culation comparison and integration into a laserscan point cloud of a built environment. *Data Brief* **2020**, *33*, 106321. [CrossRef] [PubMed]
43. Aldao, E.; González-Jorge, H.; Pérez, J.A. Metrological comparison of LiDAR and photogrammetric systems for defor-mation monitoring of aerospace parts. *Measurement* **2021**, *174*, 109037. [CrossRef]
44. Schonberger, J.L.; Frahm, J.-M. Structure-from-Motion Revisited. In Proceedings of the 2016 IEEE Conference on Computer Vision and Pattern Recognition (CVPR), Las Vegas, NV, USA, 27–30 June 2016; pp. 4104–4113.
45. Wu, C. Visualsfm: A Visual Structure from Motion System. 2011. Available online: http://ccwu.me/vsfm/ (accessed on 1 May 2021).
46. Fedorov-Davydov, G.A. Archaeological research in Central Asia of the Muslim period. *World Archaeol.* **1983**, *14*, 393–405. [CrossRef]
47. Milosz, M.; Skulimowski, S.; Kęsik, J.; Montusiewicz, J. Virtual and interactive museum of archaeological artefacts from Afrasiyab—An ancient city on the silk road. *Digit. Appl. Archaeol. Cult. Herit.* **2020**, *18*, e00155. [CrossRef]
48. 3D Digital Silk Road. Retrieved 27 January 2020. Available online: http://silkroad3d.com/ (accessed on 5 June 2021).

applied sciences

MDPI

Article

XRF and 3D Modelling on a Composite Etruscan Helmet

Joshua Emmitt [1,*], Andrew McAlister [1], Neda Bawden [2] and Jeremy Armstrong [2]

[1] School of Social Sciences, University of Auckland, Auckland 1010, New Zealand; andrew.mcalister@auckland.ac.nz
[2] Classics and Ancient History, University of Auckland, Auckland 1010, New Zealand; nbaw854@aucklanduni.ac.nz (N.B.); js.armstrong@auckland.ac.nz (J.A.)
* Correspondence: josh.emmitt@auckland.ac.nz

Featured Application: Interpolation of pXRF assay data as a texture on a 3D model.

Abstract: The presentation of X-ray fluorescence data (XRF) assays is commonly restricted to tables or graphical representations. While the latter may sometimes be in a 3D format, they have yet to incorporate the actual objects they are from. The presentation of multiple XRF assays on a 3D model allows for more accessible presentation of data, particularly for composite objects, and aids in their interpretation. We present a method to display and interpolate assay data on 3D models using the PyVista Python package. This creates a texture of the object that displays the relative differences in elemental composition. A crested helmet from Tomb 1036 from the Casale del Fosso necropolis, Veii, Italy, is used to exemplify this method. The results of the analysis are presented and show variation in composition across the helmet, which also corresponds with macroscopic and decorrelation stretching analyses.

Keywords: bronze analysis; pXRF; photogrammetry; interpolation; PyVista; decorrelation stretching; crested helmet; Casale del Fosso

Citation: Emmitt, J.; McAlister, A.; Bawden, N.; Armstrong, J. XRF and 3D Modelling on a Composite Etruscan Helmet. *Appl. Sci.* **2021**, *11*, 8026. https://doi.org/10.3390/app11178026

Academic Editor: Marek Milosz

Received: 9 August 2021
Accepted: 24 August 2021
Published: 30 August 2021

1. Introduction

The application of photogrammetry for the purpose of creating 3D models is widely applied in archaeology and heritage studies [1,2]. Models are used for a range of purposes which have traditionally fallen under the blanket term 'visualization' but also include analytical techniques (e.g., [2–4]). Likewise, portable X-ray fluorescence (pXRF) has been increasingly deployed in a range of contexts for analysis and interpretation of materials, including bronze [5,6]. One feature of non-destructive pXRF analyses is that individual assays are taken on relatively small areas of an object's surfaces (usually a few mm^2). Thus, a set of spatially discrete data points can be generated across a single object and used to assess compositional variation (e.g., [5,7]). Further, although 3D modelling and pXRF are readily applied on their own, the potential to combine them to enhance interpretation is largely unexplored. Here, we apply photogrammetric modelling and pXRF analysis together by interpolating the results of the pXRF analysis onto the photogrammetric model. The results are also assessed with decorrelation stretching of the photogrammetry model texture to further assist in the interpretation of the manufacture of an Etruscan crested helmet.

2. Context

The crested helmet under investigation is from Tomb 1036 from the Casale del Fosso necropolis at Veii (Figure 1). The helmet is currently located in the Museo Nazionale Etrusco di Villa Giulia e Villa Poniatowski di Roma (National Etruscan Museum of the Villa Giulia and Villa Poniatowski in Rome) and was examined in January 2019. Veii, located approximately 15 km north-northwest of Rome (Figure 1), was an important community in southern Etruria, which flourished during the Italian Iron Age and was traditionally considered an early rival to Rome until its conquest in 396 BCE, according to the ancient literary evidence (Livy *Ab Urbe Condita* 5.22). Whether due to Roman conquest or not, the site went into decline around this time and remained mostly abandoned afterward, with brief periods of light occupation during the medieval period [8,9]. The urban centre of Veii is situated atop a plateau with the community's various necropoleis surrounding it (Figure 1).

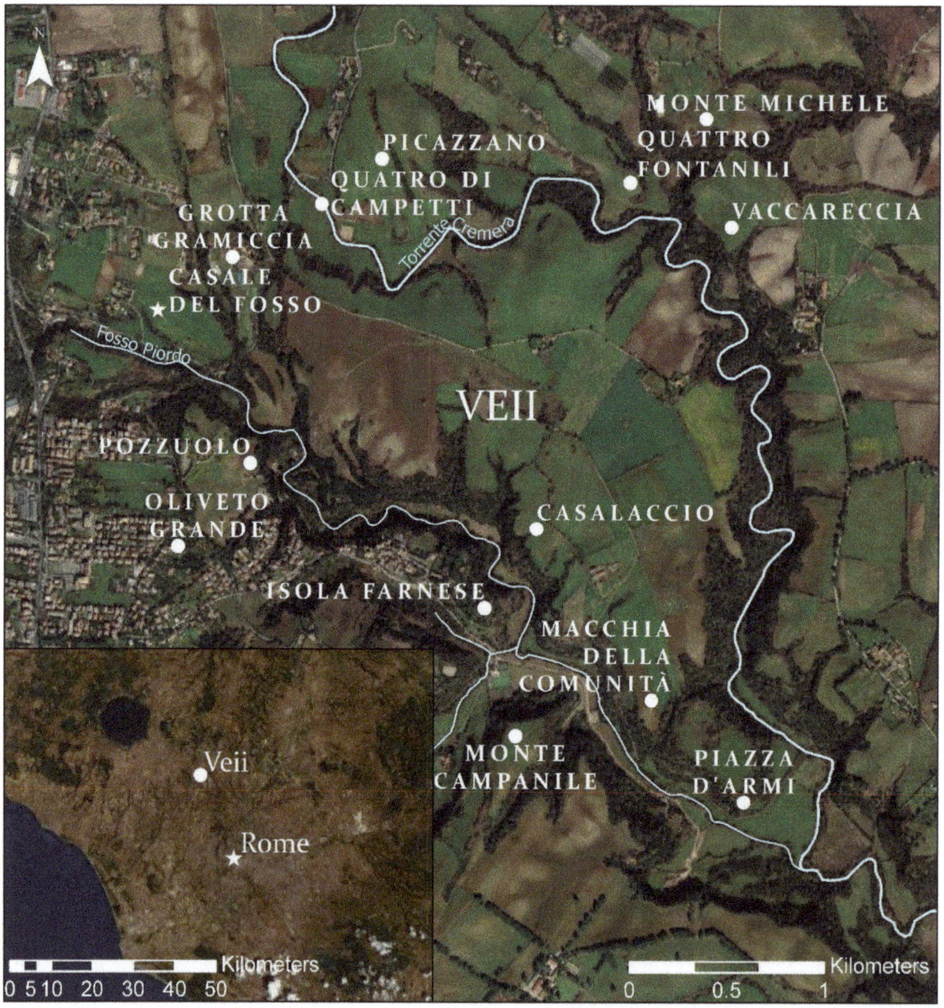

Figure 1. The site of Veii with associated necropoleis, including the location of Casale Del Fosso.

The site of Veii has been the subject of archaeological interest since the 14th century [10], although formal excavation only began in the 19th century and was initially focused on collecting valuable objects and worked marbles for private and museum collections. More systematic approaches were slowly adopted, and in the period between 1913 and 1918, a methodological investigation of several of the necropoleis was undertaken which uncovered around 1200 tombs from Grotta Gramiccia, Casale del Fosso, Pozzuolo, Monte Campanile, Valle La Fata, and Macchia della Comunità necropoleis [8,9]. Tomb 1036 from the Casale del Fosso necropolis, located to the northwest of the plateau (Figure 1), dates to between 750 and 731 BCE and was excavated in 1915. This grave was, from its initial discovery, understood to be exceptional due to both the style of burial and the number and the perceived quality of the grave goods. The context was a single inhumation burial, where the body was deposited into a *fossa*, or pit grave, in a wooden box which also contained grave goods. The grave goods included a spear, a pair of swords, a 'mace', a pair of composite 'shields', several large embossed bronze discs placed over the body, and this helmet. When excavated, the entire assemblage, including the body, the items, and the box, was encased in a thick layer of plaster and transported to the Villa Giulia in Rome to be conserved; due to a delicate conservation process, the finds were only displayed nearly a century later [11] (Figure 2).

Figure 2. Contents of tomb 1200 after the removal of plaster. After Boitani [11]: 112.

The crested helmet from Tomb 1036 is of the 'Villanovan-type' [12], which emerged in Italy during the 9th century BCE, with the first examples appearing at Tarquinia—a nearby community in south Etruria. Although possibly connected to earlier crested helmets from continental Europe, such as the Pass-Lueg helmets of the Bronze Age, and with clear connections to helmets found in San Canziano-Škocjan, Hallstatt, and Zavadintsy, these crested helmets were evidently new to Italy in this period and quickly formed an important expression of the emerging 'warrior ethos', which is argued to have dominated amongst the elite within Etruria at this time [13]. By the late 8th century BCE, when this helmet was deposited, these types of helmets seem to have taken on increasing symbolic connotations—only associated with burials of the highest status [14].

The helmet, as with others of its type, was made through an elaborate process. The two halves (front and back in Figure 3) were likely formed by 'sinking', where a flat sheet is curved by hammering into a concave indentation. The vast majority of the decoration, which was likely achieved through repoussing or punching from the inside (most notably a series of small dots, largely invisible now due to patination), would have been added during this phase as well, as it would have been impossible to achieve once the helmet was assembled. Once both halves of the helmet were complete, they were placed together and connected with three rivets along the crest—one at the forehead, one at the crown, and one at the nape. Two cast plates, each with three prongs, were then attached to the forehead and the nape, adding further decorative elements and structurally binding the various pieces together. The helmet, therefore, contains at least four distinct pieces of bronze (the two halves and the two cast plates for the forehead and the nape) and at least three rivets; the cast plates on the forehead and the nape likely contain rivets as well, although this is uncertain. Finally, an organic band (leather or textile) would have been wrapped around the lower section of the helmet.

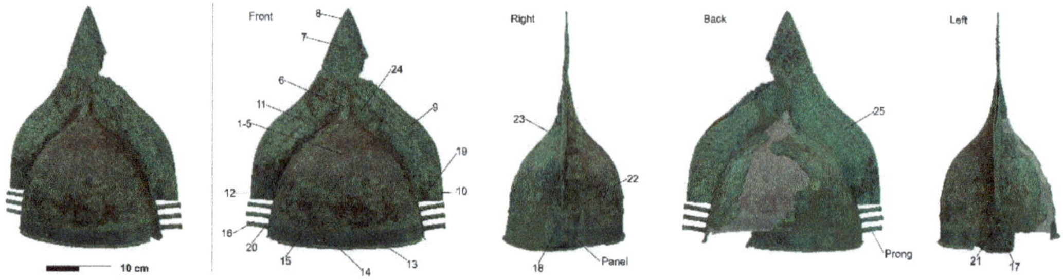

Figure 3. Photogrammetry model with the different sides of the helmet shown and location of the pXRF assays on VG015 (numbers).

While the basic construction of the helmet is evident from a visual examination, the specific details of its production are largely unexplored. It is clear that the majority of the helmet was planned out in detail before construction began. However, at present, we do not know whether the casting of the blank sheets for the two halves of the helmet, the plates for the forehead and the nape, and the rivets occurred using the same alloy. If they are all the same alloy, it suggests the entire item was produced in a single workshop where there were at least two different types of bronze work occurring, hammering and casting, which involve two distinct skill sets and sets of resources. Alternatively, if the pieces are composed of slightly different alloys, that would support the possibility that the pieces were cast and created separately. This does not preclude them all being produced in the same workshop, although it would suggest a different chaîne opératoire—and indeed one which might involve a more complex set and sequence of relationships.

All of this obviously has implications for both the role and the production of bronze items in southern Etruria. 'Villanovan-type' crested helmets were transregional artefacts which, as Iaia suggested, served as important 'vectors of technological and cultural transmission beyond their local region'—being found as far east as Fermo on the Adriatic coast and as far south as Capua and Sala Consilina in Campania and in the Piemonte region in the north [14]. It is also worth noting that southern Etruria, where this item was found, is not a mineral-rich area, and it is likely that the copper for it came from northern Etruria, in the so called 'Etruria Mineraria' (especially the Colline Metallifere north of Grosseto). Further, and as noted above, these helmets can be plausibly connected to both earlier crested helmets and contemporary finds from Europe, hinting at a vast social, cultural, economic, and technological network. Understanding how the specific production of an individual piece operated within this network can give vital clues about the nature of Etruscan society in this period.

3. Analyses

The combination of 3D data with methods which have been traditionally considered in isolation is becoming more common. Ultrasonic pulse velocity (UPV) testing and near-infrared (NIR) imaging have been combined with 3D model to provide alternative visualization techniques to assess preservation in stone objects [15,16]. In other cases, more data are being extracted from 3D models themselves. For instance, Pfeuffer and colleagues used 3D laser scanning to measure surface roughness on stone sculpture as a way to inform on suitable preservation measures [17]. In another example, as part of the wider 'Blood and Money' project this paper is part of, the 3D thickness of a different helmet was measured as a way to assess manufacturing techniques [3]. This latter method, unfortunately, is not possible on the crested helmet discussed here, as it was not logistically practical to capture an accurate 3D model of its interior surface.

To investigate the general construction techniques of the crested helmet, several methods of analysis were employed—primarily, portable X-ray fluorescence (pXRF) to characterize the elemental composition and decorrelation stretching to show potential differences in colour. The results of the pXRF analysis were plotted onto a photogrammetric model of the helmet and interpolated, which forms another level of analysis. The interpolation of pXRF data as a texture for the 3D model allows for a more integrated approach to the analysis of composite objects than is commonly considered, where composite data are often presented visually separate from the object they are from (e.g., [18]). Therefore, interpolation of pXRF data provides an alternative way of visualizing results. Finally, decorrelation stretching was undertaken on the 3D model texture to potentially provide another means of identifying the various components of the helmet.

3.1. Photogrammetry

Following Emmitt et al. [3], the setup for the photogrammetry consisted of a light tent surrounded by five LED lights on tripods, a Bluetooth controlled turntable with a custom foam stand placed over it, and photogrammetry targets printed from Agisoft Metashape 1.5.3 [19]. Twenty-five images were taken around the object per rotation, with nine rotations done at different angles for a total of 225 photos of the helmet plus one image of the background to use as a mask during model construction. A Canon EOS 7D with a Canon EF 50m f/2.5 Macro lens was used for the photographs. Colour correction was done with an X-Rite Color Checker Photo Passport 2 [20,21]. The colour checker was photographed at the start of each set-up. Raw photographs were processed as digital negatives (DNG) in ColorChecker Passport and Adobe Lightroom Classic CC with the ColorChecker Passport plugin and JPG files for photogrammetry. Agisoft Metashape 1.5.3 [19] was used for the model creation. Blank images of the workspace were used to a create masks of the helmet. The constructed sparse point cloud was edited with the following parameters: reprojection error: 0.2 px; reconstruction uncertainty: 15 px; image count: 2; projection accuracy: 2.5 px. Editing of the dense cloud was done manually with some automation based on colour where possible. A mesh was generated over the object, and a texture was created. Where required, a reflexive method of deleting stray points on the dense cloud and re-generating the mesh and the texture was conducted. The final photogrammetry model is available as part of the dataset of Emmitt and colleagues [22] (Figure 3). Due to logistical constraints, only the exterior of the helmet was modelled, thus thickness analyses, such as those done in Emmitt and colleagues [3], were not possible.

3.2. pXRF

Following Emmitt et al. [6], we used a Bruker Tracer III-SD portable X-ray Fluorescence (pXRF) analyser (Bruker, Billerica, MA, USA; also see Appendix A). The instrument employs an X-ray tube with an Rh target and a 10 mm^2 silicon drift detector (SDD) with a typical resolution of 145 eV at 100,000 cps. For analysis of copper-based alloys, we found that operating the X-ray tube with a setting of 40 keV at 5.0 μA in an air-path and through a window composed of 12 mil Al and 1 mil Ti filters (Bruker's Yellow filter) provided a good count rate for the elements of interest. Assays of artefacts were taken for 60 s each, and their locations were recorded on annotated photographs and transcribed to a photogrammetric model of each object. Reference standards were analysed using the same machine and settings outlined above (Appendix A). Each standard was analysed twice for 60 s, and the results were averaged.

In total, 25 assays were taken on the helmet (Figure 3). The assay locations were selected based on a subjective assessment of the object, which included the visually determined position of different physical features, such as decorative elements, rivets, and areas potentially less affected by patina than others. Assays 1–5 formed a 'cluster assay', which took the average of the results of five assays over a 1 cm diameter area (Figure 4). The aim of the cluster assay was to provide a more representative sample in an area that was considered less affected by patina formation. The order the five assays were recorded in is presented in Figure 4. The cluster assay measures variation over a relatively contained surface area, where both the core metal and the depositional conditions should be largely consistent [6]. This type of approach falls into the category of 'area mapping', as proposed by Karydas et al. [23], to ascertain mean values of elemental intensities across an area and address the issue of microscale heterogeneity common on bronze alloys. Assays were targeted at the cleanest areas possible (i.e., those with perceived low surface roughness caused by patina). Calibrated results from the analysis are presented in Table 1.

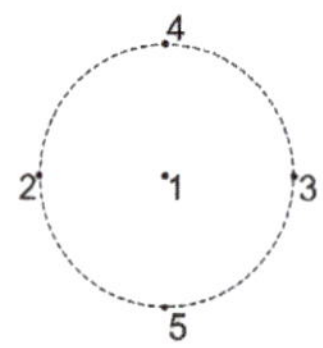

Figure 4. Example of the order in which cluster assays were taken over a 1 cm diameter circle.

In addition to the calibrated assays, the Compton scatter is also useful for analysis. The inelastic Compton scatter of the X-ray target (Rh in this case) is inversely proportional to the mean atomic weight of the sample and therefore provides a measure for the relative density of the sample area (see [24]).

Table 1. Calibrated data for assays used in the analysis. Assays 1–5 are reported as the average values of the cluster. Additionally shown are the count rates (counts per second) for the inelastic Compton scatter of the Rh target (integrated over a range of 18.5–19.5 keV).

Assay	Location	Mn %	Fe %	Ni %	Cu %	Zn %	As %	Sn %	Sb %	Pb %	Compton *cps*
1–5	Front	0.049	0.440	0.150	59.1	1.21	1.01	20.2	0.295	2.74	174.7
6	Front	0.057	0.401	0.118	55.0	1.30	0.68	17.2	0.232	1.69	172.9
7	Front	0.048	0.495	0.165	48.2	1.43	0.75	22.8	0.302	2.06	180.0
8	Front	0.048	0.442	0.138	51.9	1.48	0.76	23.3	0.329	2.15	179.6
9	Front	0.050	0.416	0.157	53.8	1.27	0.60	17.5	0.210	1.86	157.2
10	Front	0.046	0.474	0.191	42.9	1.66	0.63	22.8	0.309	1.91	157.5
11	Front	0.047	0.405	0.138	51.9	1.37	0.68	22.4	0.297	2.13	166.6
12	Front	0.051	0.524	0.195	45.4	1.72	1.30	35.8	0.441	3.79	215.1
13	Front	0.048	0.416	0.123	54.0	1.15	0.75	20.0	0.300	2.83	170.2
14	Front	0.047	0.388	0.197	42.5	1.34	0.80	20.1	0.276	2.18	154.7
15	Front	0.048	0.485	0.182	55.7	1.35	1.10	24.4	0.299	2.87	183.7
16	Prong	0.046	0.360	0.210	38.9	1.47	0.73	14.5	0.283	2.63	129.6
17	Prong	0.450	0.284	0.202	36.7	1.50	0.16	3.70	0.057	0.45	72.4
18	Panel	0.049	0.380	0.166	48.1	1.22	0.45	17.4	0.228	1.56	157.8
19	Rivet	0.050	0.601	0.167	48.3	1.56	0.79	26.3	0.366	2.19	191.2
20	Prong	0.048	0.403	0.206	42.8	1.61	1.17	21.5	0.407	3.99	162.6
21	Panel	0.047	0.309	0.185	45.0	1.29	0.77	14.2	0.212	3.01	136.6
22	Prong	0.048	0.577	0.167	47.2	1.49	0.55	19.6	0.361	2.54	155.5
23	Back	0.055	0.366	0.106	63.2	0.79	0.61	26.7	0.439	1.55	192.3
24	Front	0.045	0.408	0.159	46.8	1.46	0.50	14.5	0.178	1.09	140.9
25	Back	0.047	0.365	0.181	47.8	1.21	0.67	25.6	0.412	1.85	172.0

3.3. D Assay Interpolation

Display of the assays on the 3D model was achieved by first selecting the assay locations as points in Blender v2.90.1. This was done by manually selecting the point locations on the helmet by comparing to photographs taken during data acquisition. These points were exported as a .csv file, which included their locations relative to the rest of the model (script included with supplementary data [22]). The pXRF data were related to the point locations in the .csv format. The point values were then interpolated onto the model using the PyVista toolkit v0.28.1 [25]. The field used for interpolation could be selected from those in the .csv file thus may have been any of the pXRF results or any other desired variable. PyVista uses a Gaussian kernel interpolation, which applies the values to the area within a specified radius of the point on the mesh. The code script and the data were included in the data repository for this paper [22]. The presentation in the Juypter notebook included in the data also provided a view in 3D of the interpolated model.

In Figure 5, we display views of the helmet with interpolation values of the pXRF analysis we determined as the most useful for examining the helmet, those being Fe, Sn, Pb, As, and the Compton values. The colours for each element were designated based on the maximum value, which is red in Figure 4. As there were no elements with a measurement of zero, the colour blue was treated as a null value and was also the minimum value on the scale. The interpolations were conducted at a radius of 5 cm so as to allow some blending of values while not over-representing the assay data on the object. As the search radius was 5 cm and the helmet was much thinner than that, in some cases under 1 mm, it did not distinguish well between different sides, which has implications for the interpretation of the different sheets the helmet is made from, as we discuss further below. However, it did provide an indication of the composite nature of the helmet as it presently exists, which is the result of both the initial production techniques and the post-depositional factors—most notably patination.

The results of the interpolation demonstrate differences in the composition of the helmet. While, at first glance, it appears relatively homogenous, there are clearly differences between the crest and the bowl of the helmet and again towards the rim. Much of this difference almost certainly relates to variable patination processes, as the crest and the bowl of the helmet, on each side, are known to come from a single piece of hammered bronze. This is reinforced by the fact that the two halves of the helmet, which are made from separate pieces of bronze, seem to share broad similarities. Iron is present throughout the helmet but is higher on the rivet and towards the base of the crest. Lead is present in a higher proportion on the rivet but also along the main body of the helmet. The Compton results further demonstrate the relative homogeneity of the helmet with the exception of the prongs visible on the left view in Figure 5, which display a lower proportion of the elements displayed and a relatively lower density.

3.4. Decorrelation Stretch

The final method of analysis applied to the helmet was decorrelation stretching through the use of the DStretch [26,27] plugin for the image processing software Image-J [28]. Decorrelation stretching exaggerates the colour separation of an image and was selected as it could be applied to already available data, mainly the texture generated for the photogrammetry model. The colorspace algorithms were selected based on those that displayed the most visual distinction (Figure 6). The most distinct variation was identified at the darker end of the hue spectrum, which was also the case for pigments in rock art [29]. Distinctions were identified over some parts of the model. For instance, the colorspace LABI identified one side as more heavily patinated than the other, with the patina showing up in white. This likely relates to the way it was deposited in the tomb and later preserved in plaster, lying on its side, and also likely contributed to the differential preservation of the helmet, with one side (Back view in Figure 6) featuring significantly more damage than the other.

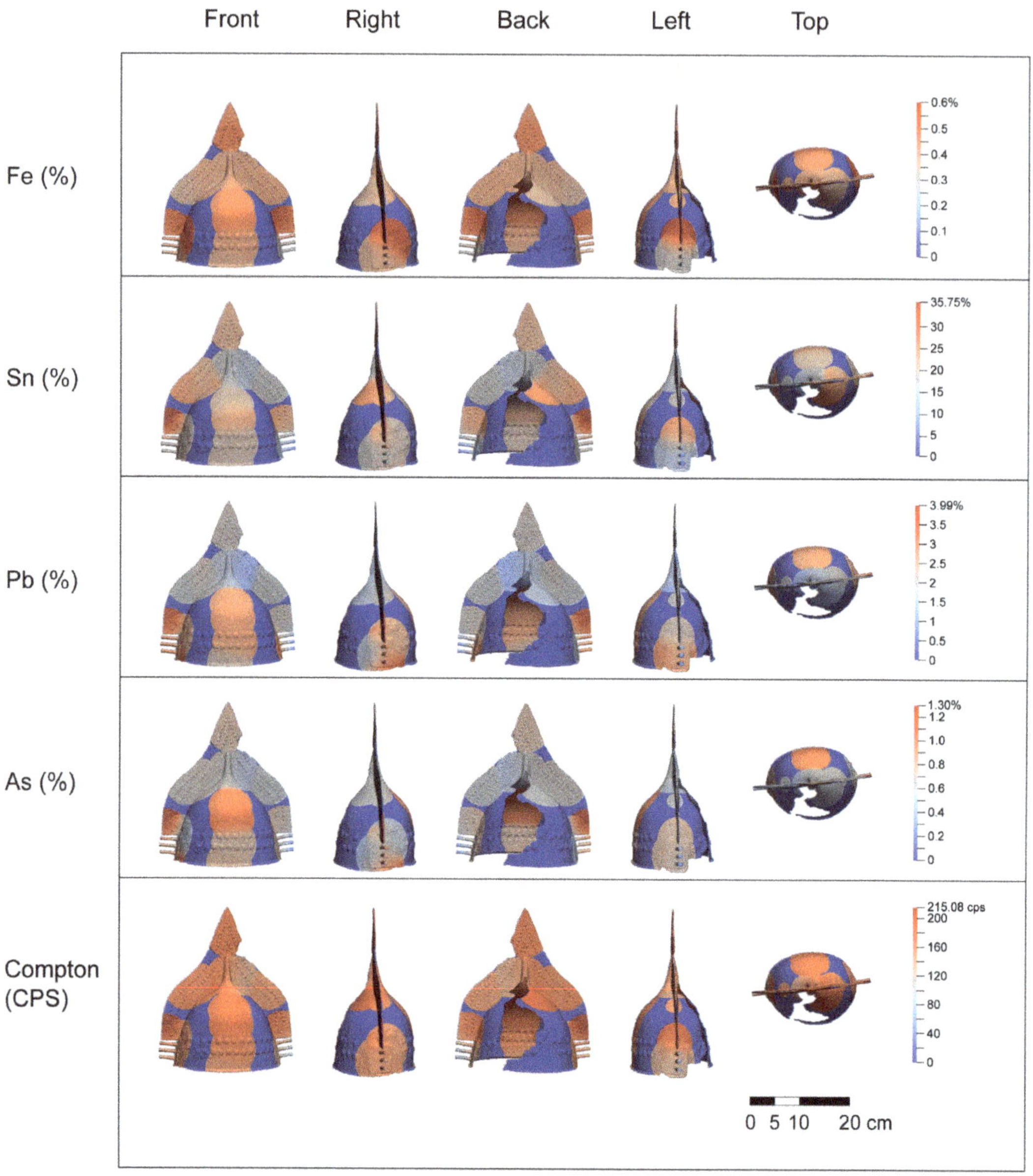

Figure 5. Interpolated values of the pXRF analyses onto the photogrammetric model.

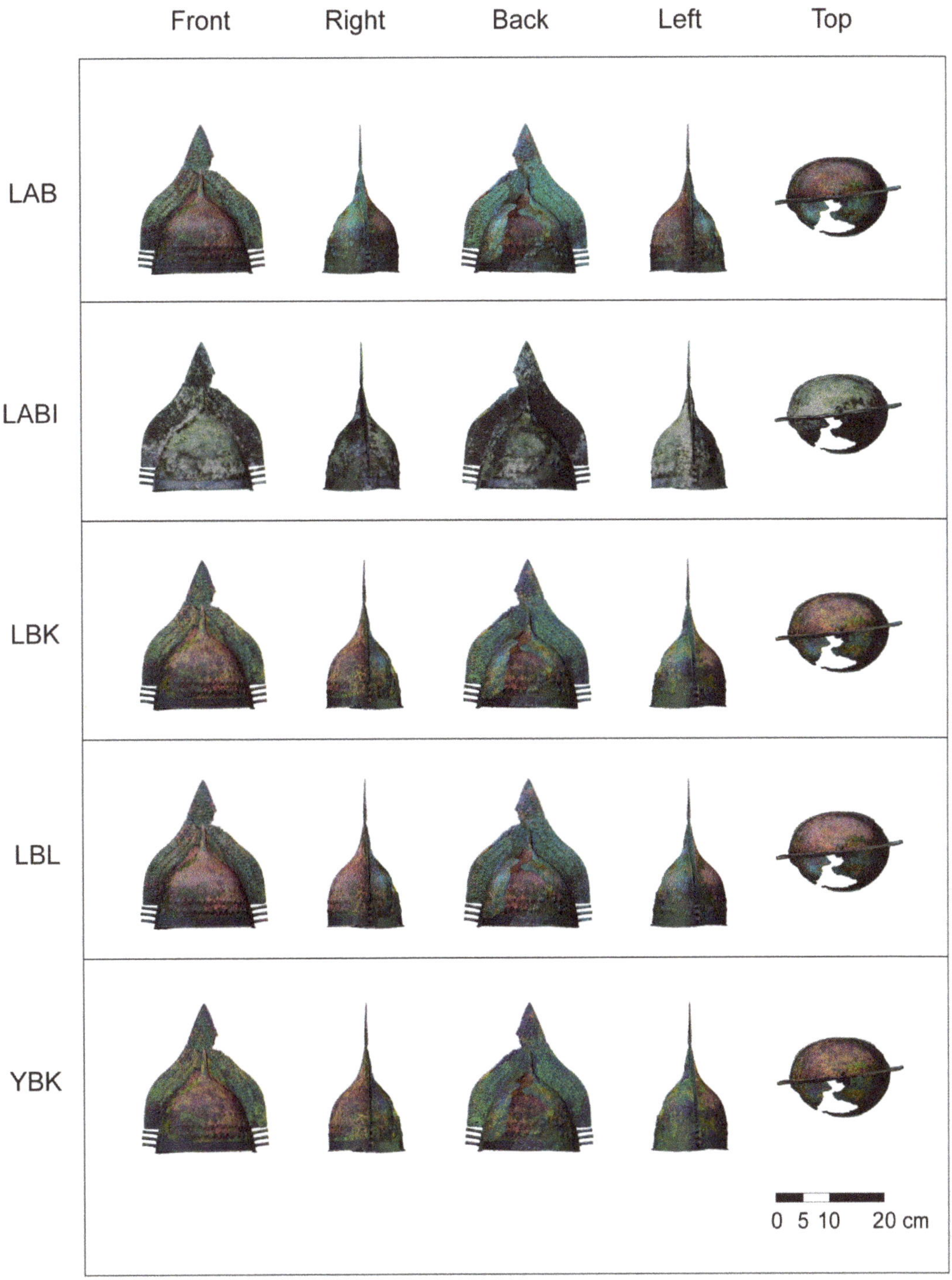

Figure 6. Results of the decorrelation stretch analysis by colorspace.

Additionally, the decorrelation stretching highlighted a band around the lower segment of the helmet, which seemed to be covered in an organic element that disappeared. This band is visible to the naked eye (Figure 7a), although it lacks definition. The decorrelation stretching clearly revealed that the band originally went all the way around the helmet, even covering the plates at the forehead and the nape (Figure 7b). The fact that the band seemed to cover the plates, which included circular design elements (one of which may have been associated with a rivet), hints that it was a later addition and was perhaps not planned at the outset (Figure 7b). It should also be noted that space for a band was left on the bowl of the helmet, which was otherwise richly decorated with repoussed and punched features (Figure 7c). This reveals details about how the item would have looked in antiquity, visibly combining organic and metal elements on the exterior in addition to the organic elements which were likely present on the inside and the chinstrap.

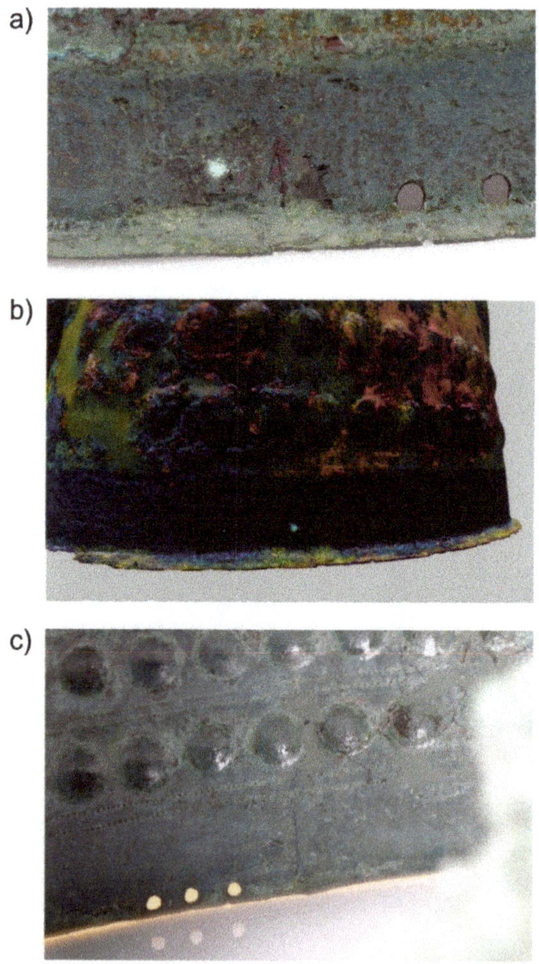

Figure 7. (**a**) Banding visible in the pantina near the rim of the helmet (darker green area). (**b**) DStrech (LBK) view of the banding (in dark blue). (**c**) Detail of intricate decoration visible from the interior of the helmet.

4. Discussion

When viewed on the interpolated surface, the pXRF data presented above suggest that the entire helmet was produced from a single type of high-tin bronze alloy. Key differences in the measurements are either within a single piece of metal or cross the boundaries between them, indicating they are due to patination, other post-depositional processes, variable thickness of the piece (visible in the Compton value), or the natural variation present in ancient bronze. Thus, we can say that all four of the main pieces of bronze (the two halves as well as the forehead and the nape plates) are of a consistent composition. This, when combined with the consistency and the planning visible in the decoration, indicates that the helmet was most likely the product of a single workshop and was produced in what can be considered a relative 'single event'. This is significant, as it indicates the presence of a complex workshop in the 8th century BCE able to handle large pours of metal into various shapes (sheets and more complex lost wax forms) as well as to perform the skilled hammerwork needed to finish the helmet.

The variations in the patina of the item are also useful. With regard to the ancient item itself, the band of discoloration likely associated with a (now lost) organic band first reveals the presence of a decorative element which was previously unidentified. Further, as the band seemed to cover elements of cast decoration, it is possible that, while the basics of the helmet's decorative scheme were established in the initial casting phase, changes could be made later—perhaps in a different/associated facet of the chaîne opératoire associated with organic materials.

With regard to the item's post-depositional history, the variations highlighted by the decorrelation stretching hint at how each side of the helmet was exposed to different conditions, which resulted in different patinas. The damaged side (back in Figures 3 and 5) was likely the one left exposed upwards during deposition, with the more patinated side (front) remaining relatively protected but also exposed to patination forming processes over time. This was also confirmed by the position of the helmet in the tomb (Figure 2).

Such interpretations as those made here are possible by examining the assay data and comparing them to photographs or a 3D model of the helmet. However, the 3D interpolation utilized here offers a replicable means of visualizing these interpretations. It also provides a more complete view of the helmet. As we discussed elsewhere [6], it is vitally important that such items should be examined as more than single data points in wider assemblages and rather as objects of potential variation in themselves. The ability to position and interpolate other data sets on 3D models creates a more holistic approach to their interpretation.

Future applications of this method should focus on a systematic collection of data across an object. As the data used here were collected during a relatively short collection window as part of a larger project, it was only possible to collect a limited number of points. Ideally, the number of points should be spaced across the object and not necessarily in relation to the different parts of it. This requires a method for collecting data which has the pXRF locations marked as part of the image texture at acquisition as opposed to marking photographs of the object as assays are taken, which we admit creates a larger margin for error than may otherwise be the case.

5. Conclusions

With exception to the photogrammetry software used (for which there are open-source alternatives), the methods employed here utilize open-source toolkits for preparation, analysis, and presentation of the 3D data. These data are published in an open repository where they can be downloaded and viewed. The technique discussed also enables others to re-run the interpolation tools we used in the paper and replicate the results of the analysis.

The results here highlight the need to consider multiple lines of evidence when examining composite objects in general, and bronzes in particular, and also the benefits of considering assays holistically and in a 3D context. Alone, the pXRF results can be presented on a table, and conclusions, such as those reached here, could have been achieved without the 3D interpolation. However, we would argue that such results are enhanced and rendered more accessible through their presentation on a 3D model. Most notably, the relationship between the points assayed as well as their relative positions on the artefact are often missed or can be misinterpreted when presented as simple tabular data. For instance, when considering the helmet discussed in this paper, the evident variability in the readings can be contrasted with the visible structuring of the helmet, its two halves and two plates, making analysis and interpretation much easier. Variability within and across different pieces is quickly and clearly evident, revealing that the most substantial variations likely relate to patination and not production. Comparisons using the Compton scatter measurements, both amongst the readings themselves and across calibrated elements, also highlight areas where, particularly on an item as thin as the helmet (c. 0.5 mm in places), additional care should be taken in interpretation due to the increased density which, at least in this instance, likely relates to thickness.

A particular benefit to the interpolation approach we propose is that it provides a visual aspect to the data which integrates them with their parent object. It also enables them to be readily compared to other visual analyses. For example, when compared against the DStretch models, another line of evidence is provided by which to interpret both the assays and the wider artefact. The photogrammetric model itself is used as a base in these cases as opposed to being limited to only a visualisation element. As 3D data and publishing become more integrated and complex (see [30]) so too do the possibilities for combining techniques that have traditionally not been considered complementary.

Author Contributions: Conceptualization, J.E. and J.A.; Methodology, J.E., A.M. and J.A.; Software, J.E.; Validation, J.E., A.M. and J.A.; Formal Analysis, J.E., A.M.; Investigation, J.E.; Resources, J.E. and J.A.; Data Curation, J.E.; Writing—Original Draft Preparation, J.E., A.M., N.B., J.A.; Writing—Review & Editing, J.E., A.M., N.B. and J.A.; Visualization, J.E.; Supervision, J.A.; Project Administration, J.A.; Funding Acquisition, J.A. All authors have read and agreed to the published version of the manuscript.

Funding: This study was funded by the Royal Society of New Zealand Marsden Fund project 'Blood and Money: The "Military Industrial Complex" of Archaic Central Italy' [17-UOA-136] awarded to Jeremy Armstrong.

Institutional Review Board Statement: Not applicable.

Informed Consent Statement: Not applicable.

Data Availability Statement: The calibrated assays used in this study are presented in Table 1. Photogrammetric models, decorrelation stretching textures, calibrated assays, and assay locations are available at: Emmitt, J., McAlister, A., Bawden, N. and Armstrong, J. (2021c) '3D and assay data published in 'XRF and 3D modelling on a composite Etruscan helmet". Zenodo doi:10.5281/zenodo.5160531.

Acknowledgments: Thanks to Rebecca Phillipps for comments on this paper, Sina Masoud-Ansari for advice on implementing the scripts, and Seline McNamee for assistance with the preparation of the figures. Thanks is due to the director and the curators from the Museo Nazionale Etrusco di Villa Giulia e Villa Poniatowski di Roma for facilitating access to their collections, permitting publication of the resultant models and data, and allocating staff, space, and time to help us during our visit. In particular, we would like to thank Giulia Bison and Valentino Nizzo for their help. The work would also have been impossible without Jono Grose and Nic Harrison of Redoubt Forge, who operated the photogrammetry workstation and assisted with the macro photography.

Conflicts of Interest: The authors declare no conflict of interest.

Appendix A. Portable X-ray Fluorescence Calibration Procedures

In this study, we used a Bruker Tracer III-SD portable X-ray Fluorescence (pXRF) analyser. The instrument employs an X-ray tube with an Rh target and a 10 mm^2 silicon drift detector (SDD) with a typical resolution of 145 eV at 100,000 cps. For analysis of copper-based alloys, we found that operating the X-ray tube with a setting of 40 keV at 5.0 μA in an air-path and through a window composed of 12 mil Al and 1 mil Ti filters

(Bruker's Yellow filter) provided a good count rate for the elements of interest. Assays of artefacts were taken for 60 s each, which was sufficient time to obtain stable results and was in line with other researchers using similar instruments [5,31,32]. Nine modern copper-based alloy reference standards were used for the calibration of these data (Table A1). Each reference standard was analysed twice, and the results were averaged.

Given the relatively small number of standards available, the calibration procedure was kept as simple as possible. Linear regressions on the net characteristic element peaks (normalised to counts-per-second) were employed, and corrections for interference peaks were included in the regression formulas where necessary (Table A2, Figure A1). For example, the escape Kα peak of Cu (6.306keV) overlaps with the characteristic Kα peak of Fe (6.405 keV) and introduces error if not accounted for in the regression equation.

In the case of three elements (Ni, Sb, and Pb), slightly negative values were occasionally obtained from the calibration. Examination of typical spectra indicates that, at low concentrations, the peaks for these elements are swamped by the tails of the much larger Cu, Kα, and Kβ peaks at low concentrations (Figure A2).

Table A1. Reference standards used for the calibration with given values (%). Uncertified values are underlined. [1] Outlier removed from calibration.

	Mn	Fe	Ni	Cu	Zn	As	Sn	Sb	Pb
				Given values					
BCR-691-A	0.20	0.20	0.10	76.50	6.02	0.19	7.16	0.50	7.90
BCR-691-B	0.40	0.50 [1]	0.20	81.00	14.80	0.10	2.06	-	0.39
BCR-691-C	0.20	0.20	-	95.00	0.06	4.60	0.20	0.50	0.18
BCR-691-D	0.10	0.10	0.30	80.50	0.15	0.29	10.10	0.30	9.20 [1]
BCR-691-E	0.30	0.30	0.50	91.50	0.16	0.19	7.00	0.70	0.20
MBH-32X PB11	0.04	0.37	0.72	92.09	1.60	0.19	3.20	0.47	1.08
MBH-32X SN6B	0.09	0.38	0.30	85.73	2.00	0.80	6.78	0.30	1.64
MBH-33X 54400A	-	0.07	0.24	86.79	3.87	0.02	3.97	0.04	4.69
MBH-33X RB2 B	0.08	0.50	0.33	82.02	9.01	0.04	4.65	0.05	2.99
				Calibrated values					
BCR-691-A	0.17	0.21	0.07	78.21	5.72	0.19	6.72	0.46	8.00
BCR-691-B	0.42	-	0.25	85.34	15.10	0.11	2.04	-	0.43
BCR-691-C	0.21	0.20	-	96.21	−0.01	4.62	0.10	0.54	0.25
BCR-691-D	0.10	0.11	0.29	79.50	0.46	0.30	9.79	0.27	-
BCR-691-E	0.27	0.32	0.48	86.63	0.31	0.19	7.40	0.68	0.22
MBH-32X PB11	0.09	0.37	0.71	91.33	1.56	0.19	3.22	0.49	1.06
MBH-32X SN6B	0.11	0.34	0.34	83.00	2.00	0.72	7.14	0.34	1.75
MBH-33X 54400A	-	0.09	0.22	86.89	3.86	0.07	3.97	0.04	4.59
MBH-33X RB2 B	0.06	0.52	0.35	84.04	8.68	0.07	4.77	0.05	2.80

Table A2. Calibration parameters. [1] Detection limit calculated as 3.3 * (σ/S).

Element	Conc. Range (%)	R^2	RMS Error (%)	Mean Abs. Error (%)	Det. Lim. [1]	Element Peak	Interference Corrections
Mn	0.04–0.40	0.944	0.03	0.02	0.05	Mn Kα1	-
Fe	0.07–0.50	0.989	0.02	0.01	0.06	Fe Kα1	Cu Kα1 (Escape peak)
Ni	0.10–0.72	0.976	0.03	0.03	0.05	Ni Kα1	Cu Kα1 (Peak overlap)
Cu	76.50–95.00	0.806	2.59	2.08	34.84	Cu Kα1	-
Zn	0.06–14.80	0.998	0.21	0.17	0.24	Zn Kα1	Cu Kβ1 (Peak overlap)
As	0.02–4.60	0.999	0.04	0.02	0.04	As Kβ1	-
Sn	0.20–10.10	0.992	0.26	0.20	0.44	Sn Kα1	-
Sb	0.01–0.70	0.987	0.03	0.02	0.06	Sb Kα1	-
Pb	0.18–7.90	0.999	0.10	0.08	0.12	Pb Lβ1	-

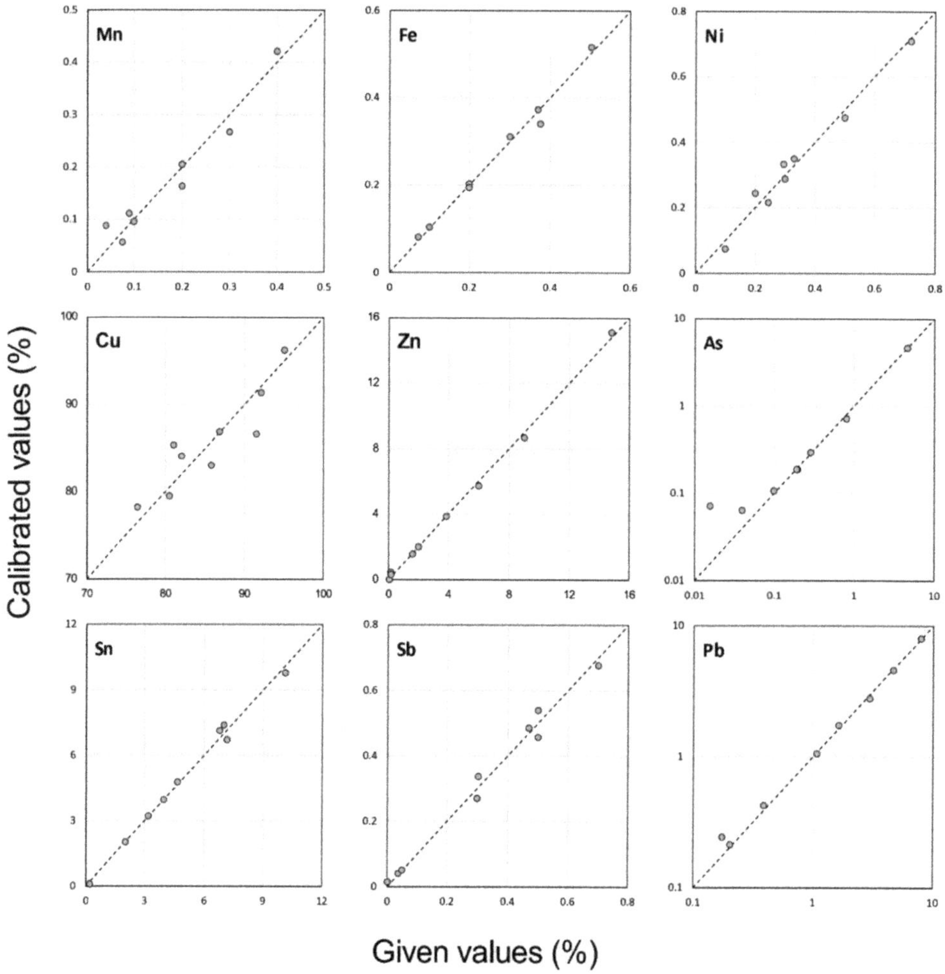

Figure A1. Scatterplots of given versus calibrated values for the reference standards. The dashed black lines show the ideal 1:1 x-y line. As and Pb are shown with logarithmic axes.

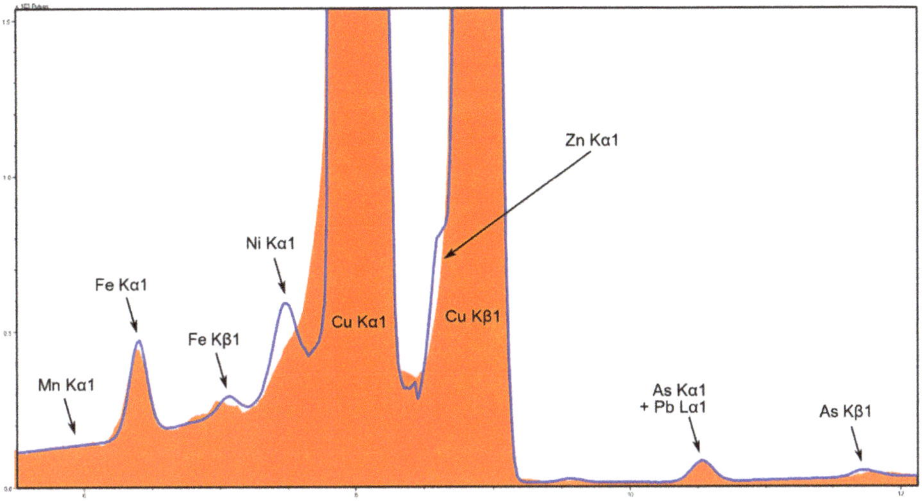

Figure A2. Spectrum of a typical bronze piece, showing the Cu Kα peak overlap on Ni and the Cu Kβ overlap on Zn. The spectrum is shown in orange, guy and the fitted peaks (using Bruker's Artax v.8.0 software) are shown as a blue line.

References

1. Porter, S.T.; Huber, N.; Hoyer, C.; Floss, H. Portable and low-cost solutions to the imaging of Paleolithic art objects: A comparison of photogrammetry and reflectance transformation imaging. *J. Archaeol. Sci. Rep.* **2016**, *10*, 859–863. [CrossRef]
2. Marín-Buzón, C.; Pérez-Romero, A.; López-Castro, J.L.; Jerbania, I.B.; Manzano-Agugliaro, F. Photogrammetry as a New Scientific Tool in Archaeology: Worldwide Research Trends. *Sustainability* **2021**, *13*, 5319. [CrossRef]
3. Emmitt, J.J.; Mackrell, T.; Armstrong, J. Digital Modelling in Museum and Private Collections: A Case Study on Early Italic Armour. *J. Comput. Appl. Archaeol.* **2021**, *4*, 63–78. [CrossRef]
4. Secci, M.; Demesticha, S.; Jimenez, C.; Papadopoulou, C.; Katsouri, I. A Living Shipwreck: An integrated three-dimensional analysis for the understanding of site formation processes in archaeological shipwreck sites. *J. Archaeol. Sci. Rep.* **2021**, *35*, 102731. [CrossRef]
5. Charalambous, A.; Kassianidou, V.; Papassavas, G. A compositional study of Cypriot bronzes dating to the Early Iron Age using portable X-ray fluorescence spectrometry (pXRF). *J. Archaeol. Sci.* **2014**, *46*, 205–216. [CrossRef]
6. Emmitt, J.; McAlister, A.; Armstrong, J. Pitfalls and Possibilities of Patinated Bronze: The Analysis of Pre-Roman Italian Armour Using pXRF. *Minerals* **2021**, *11*, 697. [CrossRef]
7. Oudbashi, O.; Davami, P. Metallography and microstructure interpretation of some archaeological tin bronze vessels from Iran. *Mater. Charact.* **2014**, *97*, 74–82. [CrossRef]
8. Colini, A.M. Veio. Scavi nell'area della città e della necropolis. *Not. Degli Scavi* **1919**, *1919*, 3–12.
9. Ward-Perkins, J.B. Veii: The historical topography of the ancient city. *Pap. Br. Sch. Rome* **1961**, *19*, 1–124. [CrossRef]
10. Delpino, F. Cronache Veientane. In *Storia Delle Ricerche Archeologiche a Veio I. Dal XIV Alla Meta del XIX Secolo*; Consiglio Nazionale delle Ricerche: Rome, Italy, 1985.
11. Boitani, F. Veio-Casale del Fosso, tomba 1036. In *Veio, Cerveteri e Vulci. Città d'Etruria a Confront*; Moretti Sgubini, A.M., Ed.; Catalogo della Mostra: Rome, Italy, 2001; p. 112.
12. Hencken, H. *The Earliest European Bronze Helmets: Bronze Age and Early Iron Age*; Peabody Museum Press: Cambridge, MA, USA, 1971.
13. Iaia, C. Produzioni toreutiche della prima età del ferro in Italia centro-settentrionale. In *Stili Decorativi, Circolazione, Significato*; Biblioteca di Studi Etruschi: Rome, Italy, 2005.
14. Iaia, C. Bronzesmiths and the Construction of Material Identity in Central Italy, (1000–700 BCE). In *Crafting Connections: Production, Trade, and Identity in Pre-Roman Italy 900-400BCE*; Armstrong, J., Cohen, S., Eds.; Routledge: London, UK, in press.
15. Akoglu, K.G.; Kotoula, E.; Simon, S. Combined use of ultrasonic pulse velocity (UPV) testing and digital technologies: A model for long-term condition monitoring memorials in historic Grove Street Cemetery, New Haven. *J. Cult. Herit.* **2020**, *41*, 84–95. [CrossRef]
16. Adamopoulos, E.; Rinaudo, F. Near-infrared modeling and enhanced visualization, as a novel approach for 3D decay mapping of stone sculptures. *Archaeol. Anthropol. Sci.* **2020**, *12*, 138. [CrossRef]

17. Pfeuffer, C.; Rahrig, M.; Snethlage, R.; Drewello, R. 3D mapping as a tool for the planning of preservation measures on sculptures made of natural stone. *Environ. Earth Sci.* **2018**, *77*, 312. [CrossRef]
18. Mozgai, V.; Bajnóczi, B.; May, Z.; Mráv, Z. Non-destructive handheld XRF study of archaeological composite silver objects—The case study of the late Roman Seuso Treasure. *Archaeol. Anthropol. Sci.* **2021**, *13*, 83. [CrossRef]
19. Agisoft. Agisoft Metashape Professional, 1.5.3. 2019. Available online: http://www.agisoft.com/downloads/installer/ (accessed on 8 August 2021).
20. Vitorino, T.; Casini, A.; Cucci, C.; Gebejesje, A.; Hiltunen, J.; Hauta-Kasari, M.; Picollo, M.; Stefani, L. Accuracy in Colour Reproduction: Using a ColorChecker Chart to Assess the Usefulness and Comparability of Data Acquired with Two Hyper-Spectral Systems. In Proceedings of the Computational Color Imaging, Cham, Germany, 24 March 2015; pp. 225–235.
21. Marziali, S.; Dionisio, G. Photogrammetry and macro photography. The experience of the MUSINT II Project in the 3D digitizing process of small size archaeological artifacts. *Stud. Digit. Herit.* **2017**, *1*, 298–309. [CrossRef]
22. Emmitt, J.; McAlister, A.; Bawden, N.; Armstrong, J. *3D and Assay Data Published in "XRF and 3D Modelling on a Composite Etruscan Helmet"*; Zenodo: Geneva, Switzerland, 2021. [CrossRef]
23. Karydas, A.G.; Anglos, D.; Harith, M.A. Mobile Micro-XRF and LIBS Spectrometers for Diagnostic Micro-Analysis of Ancient Metal Objects. In *Metals and Museums in the Mediterranean: Protecting, Preserving and Interpreting*; Argyropoulos, V., Ed.; Department of Conservation of Antiquities and Works of Art: Athens, Greece, 2008; pp. 141–177.
24. Franzini, M.; Leoni, L.; Saitta, M. Determination of the X-ray mass absorption coefficient by measurement of the intensity of Ag Kα compton scattered radiation. *X-Ray Spectrom.* **1976**, *5*, 84–87. [CrossRef]
25. Sullivan, C.B.; Kaszynski, A.A. PyVista: 3D plotting and mesh analysis through a streamlined interface for the Visualization Toolkit (VTK). *J. Open Source Softw.* **2019**, *4*, 1450. [CrossRef]
26. Harman, J. Using decorrelation stretch to enhance rock art images. In Proceedings of the American Rock Art Research Association Annual Meeting, Sparks, NV, USA, 28 May 2005.
27. Harman, J. Using DStretch for Rock Art Recording. *Int. Newsl. Rock Art* **2015**, *72*, 24–30.
28. Schneider, C.A.; Rasband, W.S.; Eliceiri, K.W. NIH Image to ImageJ: 25 years of image analysis. *Nat. Methods* **2012**, *9*, 671–675. [CrossRef]
29. Palomar-Vázquez, J.; Baselga Moreno, S.; Viñals Blasco, M.-J.; García-Sales, C.; SanchoEspinós, I. Application of a combination of digital image processing and 3D visualization of graffiti in heritage conservation. *J. Archaeol. Sci. Rep.* **2017**, *12*, 32–42. [CrossRef]
30. Schreibman, S.; Papadopoulos, C. Textuality in 3D: Three-dimensional (re)constructions as digital scholarly editions. *Int. J. Digit. Humanit.* **2019**, *1*, 221–233. [CrossRef]
31. Roxburgh, M.A.; Heeren, S.; Huisman, D.J.; Van Os, B.J.H. Non-Destructive Survey of Early Roman Copper-Alloy Brooches using Portable X-ray Fluorescence Spectrometry. *Archaeometry* **2019**, *61*, 55–69. [CrossRef]
32. Zararsiz, A.; Zimmermann, T. The Missing Jigsaw Piece–pXRF Bulk Analysis of the Karaburun Dagger and Some General Considerations on Metalwork in Early Bronze Age Western Anatolia. *Cedrus* **2020**, *8*, 65–74. [CrossRef]

applied sciences

MDPI

Article

Scanning of Historical Clothes Using 3D Scanners: Comparison of Goals, Tools, and Methods

Kamil Żyła [1,*], Jacek Kęsik [1], Filipe Santos [2] and Ginevra House [3]

[1] Department of Computer Science, Faculty of Electrical Engineering and Computer Science, Lublin University of Technology, Nadbystrzycka 36B, 20-618 Lublin, Poland; j.kesik@pollub.pl
[2] CI&DEI—Centre for Studies in Education and Innovation, Polytechnic of Leiria—Higher School of Education and Social Sciences, Rua Dr. João Soares, Apartado 4045, 2411-901 Leiria, Portugal; filipe.santos@ipleiria.pt
[3] Independent Researcher, York YO10 4DU, UK; ginevra.house@yahoo.co.uk
* Correspondence: k.zyla@pollub.pl

Abstract: Due to the UN and EU's strong interest in digitizing cultural heritage, the application of 3D scanning technology is gaining importance, even in the case of under-explored areas, such as the 3D scanning of historical clothes. This article discovers and compares methodologies of 3D scanning of historical clothes presented in the literature in order to determine if a new methodology is needed. PRISMA protocol was used to browse scientific sources in an organized way. We posed the following research question: How have 3D scanners been used to digitize historical clothes? The very limited number of works identified, despite our thorough search, allows us to conclude that this topic is very new, and a lot of research can be conducted in the future. We analyzed the methodologies proposed by other authors, taking into account factors such as what was scanned, what was the purpose of scanning, what hardware and software was used, how detailed the description was, etc. It was revealed that other authors explored the topic insufficiently and no complex and coherent methodology of 3D digitization of historical clothes is present. Generally, the field of 3D scanning of historical clothing remains, at this point, very small and fragmented. This work is one of steps to change it.

Keywords: 3D scanning; historical clothes; heritage digitization; methodology evaluation; systematic literature review

Citation: Żyła, K.; Kęsik, J.; Santos, F.; House, G. Scanning of Historical Clothes Using 3D Scanners: Comparison of Goals, Tools, and Methods. *Appl. Sci.* **2021**, *11*, 5588. https://doi.org/10.3390/app11125588

Academic Editor: Enrico Vezzetti

Received: 26 May 2021
Accepted: 11 June 2021
Published: 17 June 2021

1. Introduction

3D scanning in the field of cultural heritage involves computer science and cultural heritage preservation, resulting in the emergence of unique solutions. In this article we attempt to systematize and analyze the current state of knowledge about the digitization of historical clothes, which tend to be very delicate heritage artefacts, using 3D scanners.

1.1. Rationale

3D technologies are being used more widely in the area of cultural heritage each year. This holds true for museums that feel the need to make their exhibitions more attractive in the era of the digital society. Enabling access to museum collections for a wide audience by means of digital media has become relatively easy and has been successfully used not only by the world's leading museums but by smaller institutions as well [1–3].

However, the application of 3D scanning technology in digitization of cultural heritage objects is still a relatively young field; the methodologies are as yet under-developed, and knowledge about these techniques is not widespread [4]. One such under-explored area is the problem of the 3D scanning of historical clothes. Textiles are among the most perishable artefacts, but even the smallest fragments of textile are of value in understanding production technology and cultural significance [5,6].

Among museums with clothing in their collections, the most common methods of sharing resources in digital media are the presentation of 2D photographs [7], 360° photographic views [8], or 3D models made by means of very laborious CAD modeling [9] or photogrammetry. These techniques are used by a number of famous institutions, such as the Louvre in Paris, the V&A in London, the Kyoto Costume Institute in Japan, Google Arts & Culture platform, the Museum at FIT (MFIT), Drexel University's Historic Costume Collection, the Digital Clothing Center (DCC) at Seoul National University, and the Museum of London [10].

Within the academic literature, scientific studies on the topic of the 3D digitization of historical clothing tend to focus on the following areas:

- Building 3D models with CAD (Computer Aided Design) tools, either using dedicated software for clothes (such as CLO 3D, Gerber AccuMark 3D, Optitex or Lectra Modaris) or general modeling solutions (such as MeshLab, Blender 3D Creation Software, AutoDesk 3D, or Rhinoceros 3D);
- Construction of 3D models using photogrammetric techniques (3D model generation based on a series of photos of the object).

The above techniques are of limited accuracy when it comes to reproducing the shape and structure of garments. Greater precision and accuracy can be obtained using 3D scanners; however, only a small number of studies mention the use of 3D scanning in this process. Three-dimensional scanners, especially those based on structured light, provide a set of important features, especially with regards to obtaining a precise representation of the surface with simultaneous reproduction of its color. The value of these features is confirmed by the frequent use of this type of scanner in other aspects of the 3D digitization of cultural heritage objects. For these reasons, 3D scanners are the focus of the present study.

1.2. *Three-Dimensional Digitization Technologies*

Three-dimensional digitization of a real object can be understood as a process of translating the geometrical and visual features of the object's surface into a mathematical description. A 3D model can be perceived as a product of this process. Among the commonly used technological solutions for obtaining knowledge about these objects' features, we could mention the following:

1. CAD modeling;
2. Touch scanning;
3. Laser scanning;
4. Structured light scanning;
5. Computerized tomography scanning;
6. Photogrammetry.

CAD modeling might be perceived as a highly manual approach to producing a 3D model of a real object, in that a lot of human effort is needed to map the real object's geometrical and visual features to its 3D model. The other abovementioned solutions can be perceived as being more automated: they heavily utilize specialized algorithms that automatically recognize and reproduce the object's geometrical and visual properties in the form of a 3D model. The main disadvantage of CAD modeling, compared to the other mentioned solutions, becomes evident when the goal (and this is a very frequent one) is to obtain a 3D model offering an accurate representation of a real heritage object. More specifically, the disadvantage of CAD lies in the excess of freedom it provides in mapping the shape, such that the fidelity of the reconstruction can potentially be undermined. At the same time, striving to build a faithful model requires a significant amount of work and a high level of expertise from the creator. This type of approach is also used in heritage settings [11]; however, the resulting 3D model should be treated more as a representation of a given type of design than as a digital copy of a specific exhibit.

The touch scanning technology uses devices that have an arm with a probe that makes physical contact with the object. When the object is touched, the position of the probe is read, which allows the device to record the exact 3D position of a given point on the object's surface. If samples are collected densely enough, it is possible to recreate the shape of the object with the required accuracy. However, exposing a fragile cultural heritage object to repeated contact with a hard probe risks potential damage. Although the accuracy of the obtained surface measurements is high, scanning by touching the surface of the object so as to reproduce every detail is extremely burdensome, and in some cases, even impossible. A further disadvantage of this solution is that no information is obtained regarding the color of the scanned surface.

In laser scanning devices, the physical probe has been replaced with a laser beam. The device directs the beam to a specific point in space and captures the returning beam reflected from the object's surface. On the basis of the reflected beam's parameters, the distance to the reflection point can be calculated, indicating the position of a given surface point in space. Modern scanners also have an integrated camera that allows the colors to be read and matched at the designated 3D points. Modern scanners use Class 1 lasers operating in the infrared light band. Therefore, they do not adversely affect the surface of the scanned objects, unless it is particularly sensitive. Most often, scanners of this type provide a large scanning range and accuracy expressed in millimeters, and they are used to scan large cultural heritage objects (vehicles, buildings, and even entire historical sites) [12–16].

Structured light scanning involves devices that project an image onto the scanned object. The scanner projector displays an image containing a certain structure (most often fringes of different widths). The distortions of the fringes formed as the projection falls on the three-dimensional surface of the scanned object are captured by cameras that have been calibrated with the projector. The differences between the original and the captured image are used to calculate the topography of the object's surface in 3D. It is also possible to obtain the color of the object's surface, along with its 3D shape. Structured light scanners can be stationary (i.e., the scanning head is stationary and the scanned object is moved) or mobile (the scanning head is movable, often handheld, while the scanned object remains stationary). It is possible to scan objects of various sizes, from a few millimeters to several meters, to an accuracy of fractions of a millimeter, depending on the device.

Computerized tomography (CT) scanning technology is more widely used in medicine; in fact, most descriptions of the use of CT scanning in the area of cultural heritage indicate the use of devices intended for medical purposes. The operation involves the construction of a 3D model based on a series of X-ray images of the object. In the area of cultural heritage, it has a narrow set of applications, where the goal is not so much to obtain a model of the surface of the object but to visualize its interior using non-invasive methods.

The photogrammetry solution comes from the GIS (Geographic Information System) field, where it is used to determine the topography of a landscape based on aerial photographs. To digitize an object in this way, it is necessary to acquire a significant number of surface images (for example, photos taken with ordinary digital cameras) covering overlapping areas. Dedicated software finds easily distinguishable points in each image and then matches the equivalents in different images. On this basis, the position of these points in 3D space can be determined. The physical scale of the scanned object cannot be ascertained directly from images, thus obtaining the correct size of an object requires additional determination of its scale, for example, by means of a calibration object placed next to the scanned object. In the area of cultural heritage, photogrammetry is used in 3D modeling of both small exhibits and larger areas, such as archaeological sites or buildings. In most descriptions of scanning methods, photogrammetry is seen as a separate technique, and is not classified as 3D scanning. For the purposes of this article, it is included, but at arm's length, as it were, to highlight the ambiguous or erroneous use of the term "3D scanning" in the literature.

1.3. Objectives

Thus far, very few publications have been devoted to the scanning of historical clothes using 3D scanners. To the best of our knowledge, the full potential of 3D scanners is not being exploited in this particular field. This situation prompted us to ask whether any scientific approach to this topic has been developed. The usefulness of specific technological solutions and methods can be properly confirmed by asking whether their application is discussed in the form of scientific studies, and whether there are any academics tackling the complex question of a methodology for scanning clothing—a highly diversified category manifested in a huge variety of different shapes (from shoes to hats) and materials (from linen to golden threads reflecting scanner light). We also wished to explore whether any methodology for analyzing results has been developed.

To investigate this, we started from the following research question:

RQ: "How have 3D scanners been used to digitize historical clothes?"

To answer this question, we defined the following research objectives:

RO1: Identify whether studies on this topic exist and how many there are.

RO2: Identify the purpose driving the creation of 3D scans and the technology used.

RO3: Assess how well the literature describes the 3D scanning procedure.

2. Research Methodology

We chose to put discovery and analysis of methodologies of 3D scanning of historical clothes in the frames of well-established protocol, which is generally advised in the literature, e.g., by Reference [17]. We followed the procedure outlined in the PRISMA-P 2015 protocol, a revised and improved version of PRISMA 2009 [18] that has been widely adopted by the scientific community [19,20]. We found the protocol useful in organizing our search, and defining criteria and procedure of the methodologies evaluation and comparison.

2.1. Eligibility Criteria

Our goal was to select studies with relation to the 3D scanning of historical clothes using dedicated professional devices called 3D scanners. We did not investigate studies on the creation of 3D models from photographs taken, for example, by DSLR cameras or iPhone devices. As such, this article deliberately excludes many valuable studies about photogrammetry usage for digitizing historical clothes.

2.1.1. Inclusion Criteria

We selected studies that do the following:

1. Relate to the 3D scanning of historical clothes using 3D scanners;
2. Provide any details on the procedure of 3D scanning of clothes, or any proof (pictures, descriptions, models, etc.) of the 3D scanning of clothes.

We defined historical clothes as any kind of clothing (for humans or even for dolls) that are considered heritage items (according to the authors of the analyzed studies); are part of a private collection of historic value; or form part of a museum's collection. Photogrammetry was outside scope of our search criteria, although studies that referred to photogrammetry "as 3D scanning" were included into the analysis as an example of imprecise usage of terminology by authors. Finally, only scientific sources were included in the scope of the analysis, such as peer reviewed journals, conference proceedings, books, book chapters, and theses (undergraduate, master's, and doctoral) if returned by queries run in the chosen databases. We were able to access the full text of all studies included into the analysis, encountering no problem with paywalls. No limits were placed on publication dates in our search terms.

2.1.2. Exclusion Criteria

We excluded the following:

1. Studies containing the query keywords that were not related to the 3D scanning of historical clothes;
2. Studies about the 3D scanning of jewellery, having previously agreed that jewellery should not be considered clothing but treated as an accessory;
3. Studies that only mention that clothes are a type of heritage item that could be 3D scanned but provide no details whatsoever on procedure of scanning clothes, present no examples (pictures, descriptions, and models) of 3D-scanned clothing, and offer no background information on clothes. Knowing the background of clothes is essential to determining whether an item of clothing can be assumed to be historical.

We used English keywords in our bibliographic database queries, based on the assertion that English is the common language of modern science. As a result, we may have excluded some less visible studies written in other languages and with no English-language abstract or keywords. Some databases returned results in other languages, such as German, French or Dutch, even though the queries were in English. Thankfully, our team had sufficient skills in these languages that we could carry out the screening process on these studies.

2.2. Information Sources

The academic databases considered for this review were ACM, IEEE (two main engineering professional associations), Scopus, Web of Science, Wiley, Springer, and ScienceDirect (main publishers of academic studies). All of these databases index conference proceedings and journals that are considered reputable and of good quality.

We also expanded our query to Google Scholar to ensure a more complete coverage of the topic (accounting for variability between the indexing in each database). Using Google's academic engine has some advantages, as it detects works of lower visibility that have been archived in more informal repositories such as academic social networks (for example, ResearchGate and Academia.edu).

2.3. Search Strategy

The search was conducted on 22 February 2021. The keywords used in the query string executed for each chosen bibliographic database were derived from the research question.

The keyword summary can be found in Table 1. We decided to search for the formal synonyms of the word "clothes" according to the online Cambridge English Dictionary (https://dictionary.cambridge.org/pl/dictionary/english/clothes). The list was further expanded with other words that the authors knew from experience might occur in scientific sources in the context of the 3D scanning of clothes, namely "garment", "costume", "fashion", "textile", and "outfit". Another element of the query was technology-related—keywords which reflect the activity of 3D scanning, its results, and devices used for such purpose, i.e., 3D scanners. Finally, since this study focuses on historical clothes, the keywords included variations of the words "historic", "heritage", "museum", and "old".

Some of the keywords were quite generic and widely used in areas other than the digitization of clothes; for example, a large number of irrelevant results were generated from the fields of medicine and materials engineering. We included such terms intentionally, as we already knew that the set of relevant studies would be quite small, thus we needed our query to be broad enough that nothing relevant would be missed.

Based on the query keywords summary, the following representative query was built: ("3D scan*") AND (histor* OR heritage OR muse* OR old*) AND (clothes OR clothing OR fashion OR textile* OR costume* OR apparel* OR garment* OR dress* OR outfit* OR attire*).

Table 1. Query keywords.

Word	Reason	Wildcards
clothes, clothing, apparel, dress, attire	Formal synonyms of "clothes" (according to the Cambridge English Dictionary)	Asterisk (*)—to include, when possible, both singular and plural form
garment	A piece of clothing	
costume	A frequent synonym of historical clothes	
fashion	A style that is popular at a particular time, especially in clothes, hair, make-up, etc.	
textile	Cloth made by hand or machine in large quantities	
outfit	A set of clothes worn for a particular occasion or activity	
3D scanning	Reflects the activity of 3D scanning	Asterisk (*)—to include a variety of tense forms
3D scan	Reflects the result of 3D scanning	Asterisk (*)—to include a singular and plural form
3D scanner	Reflects the use of 3D scanner	
history	Reflects the historical nature of clothes	Asterisk (*)—to include all variations of the word
museum	Historical clothes might be managed by a museum	
heritage	Historical clothes might be called an aspect of heritage	Asterisk (*)—to include other forms of the word
old	Historical clothes might be called old	

The query assumes the following:

1. Wildcards inside of quotation marks are interpretable;
2. An asterisk (*) represents a string of zero or more characters;
3. The words inside quotation marks should be in the provided order in the text;
4. "AND" and "OR" are logical operators;
5. Parentheses group logical expressions.

The representative query was then adjusted to each of chosen databases, to meet the specific criteria of each search engine. We tried our best to ensure the same meaning of the query each time. Not all of chosen search engines provided full control over the query interpretation; for example, ScienceDirect handled variations of keywords, such as different tenses for scanning activity verbs, or plural and singular forms, but did not allow wildcards and only allowed a limited number of keywords. In case of severe limitations on numbers of keywords or use of wildcards, we split our query into many smaller ones to preserve its original coverage, removing duplicated results if necessary. Thankfully we were able to obtain comparable interpretations (or a wider interpretation, as with Google Scholar) of the representative query in each of the chosen databases, which suggests that in each case we searched for all the necessary keywords variations.

In each database we chose the search type according to the following criteria: (1) keywords+abstract+title and (2) full text search if possible. The scope of each search was set to the broadest possible, in order to ensure that all possible sources of studies were included, with no date-of-publication limits, no discipline or topic excluded, no study type excluded, and so on. The number of query results and the search type for each database used is presented in Table 2. The search type names from the table comply with the names used by particular databases.

Table 2. Query results.

Database	Search Type	No. Studies
ACM Digital Library	Anywhere	332
IEEE Xplore	All metadata and full text	405
Scopus	Title–abstract–keywords	20
Web of Science	Topic search	22
Wiley Online Library	Anywhere	558
Springer Link	Anywhere	1050
ScienceDirect	Anywhere	558
Google Scholar	Anywhere	~9770 *

* Google scholar returned information that about 9770 results were found. However, the search engine was constructed in such a way that it only displayed 1000 results returned as the "most relevant".

2.4. Study Records

2.4.1. Selection Process

We identified a total of 2945 studies that originated from the high-quality bibliographic databases (ACM Digital Library, IEEE Xplore, Scopus, Web of Science, Wiley Online Library, Springer Link, and ScienceDirect). A further 46 records were identified via Google Scholar. We identified 66 duplicates among all the records. After removing these, 2879 records underwent the screening phase in which the eligibility criteria were applied. Only 12 studies were found eligible and were included in the subsequent qualitative and quantitative analysis. For details see the PRISMA flow diagram in Figure 1.

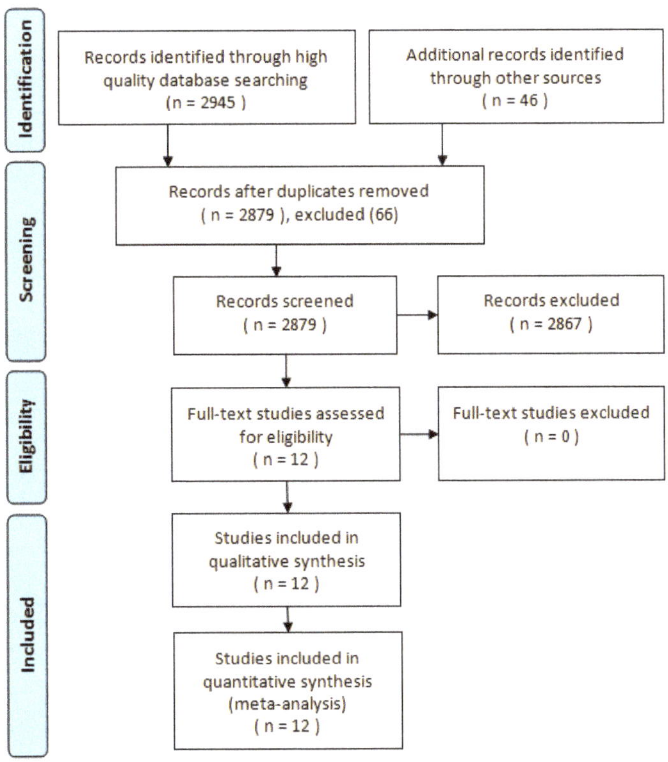

Figure 1. PRISMA flow diagram.

We decided to process the results from Google Scholar separately, as we perceive this source as less of a high-quality bibliographic database. Moreover, it presents too many significant technical problems when it comes to applying the PRISMA protocol steps in the same way as was achieved for records from other chosen databases. Google Scholar is not designed for literature reviews; some of issues, among others, include imprecise output (for example, estimated number of results, a limit of 1000 records shown); non-transparent search algorithms; and insufficient support for bibliometric analysis and deduplication. Nevertheless, it is a source of records for studies not indexed in the other databases which might nevertheless be relevant for our review. Thus, we decided to briefly analyze the records returned by Google Scholar, choosing any that looked relevant and creating a list that could be included in the PRISMA-P workflow as additional records identified through other sources. This was performed independently by two of the authors (Żyła and Kęsik).

The same two authors independently screened the titles and abstracts yielded by the search against the inclusion criteria to reduce the possibility of excluding relevant reports. Discrepancies were resolved through an inter-rater agreement. Full studies that appeared to meet the inclusion criteria were extracted, as were those where there was any uncertainty. The two reviewers then screened the full study and decided whether it met the inclusion criteria. Disagreements were solved through discussion, and the reasons for excluding studies were documented.

2.4.2. Data Management

Due to the relatively small number of studies that were eligible for inclusion in this study (n = 12), we used the more traditional form of data management, namely the online spreadsheet software Google Sheets. Because it was accessible online, this spreadsheet served as a collaborative tool, allowing the three authors to discuss and arrive at a consensus on issues of quality assessment.

In order to compute the statistics and manage duplicates, we developed our own software which took text files generated by bibliographic databases in CSV or BibTex format as an input. The metadata of studies was extracted (DOI numbers, titles, etc.) and compared accordingly. The comparisons were made in a case insensitive manner, whenever appropriate.

2.4.3. Data Collection Process

To ensure consistency between the two reviewers (Żyła and Kęsik), calibration exercises were undertaken before the review. Disagreements were solved through discussion, and one arbitrator (Santos) adjudicated on unresolved disagreements. It was decided to contact study's authors if any uncertainties remained.

2.5. Data Items

The thorough qualitative and quantitative analysis of the selected studies required a precise and uniform set of data to be extracted in each case. The kind of data that was extracted and categorized according to the particular research objective is presented in Table 3.

Table 3. Data items to be extracted from the studies.

Research Question: How Have 3D Scanners Been Used to Digitize Historical Clothes?	
Research Objective	**Data Items**
RO1: Identify whether studies on this topic exist and how many there are	• The source of study (e.g., bibliographic database name) • The authors' data • Type of publication • Keywords

Table 3. *Cont.*

<table>
<tr><th colspan="2">Research Question: How Have 3D Scanners Been Used to Digitize Historical Clothes?</th></tr>
<tr><th>Research Objective</th><th>Data Items</th></tr>
<tr><td>RO2: Identify the purpose driving the creation of 3D scans and the technology used</td><td>• Scanner type/technology type used for scanning
• Purpose of 3D scan/reason for scanning
• What was scanned</td></tr>
<tr><td>RO3: Assess how well the literature describes the 3D scanning procedure</td><td>Hardware and software (equipment): How well is the hardware and software described?
• Procedure: How well is the procedure described?
• Objects: How well are the scanned objects described?
• Guidelines: Is the paper focused on giving guidelines?</td></tr>
</table>

2.6. *Risk of Bias in Individual Studies*

The PRISMA-P protocol encourages careful thinking in anticipation of possible bias (or risk of bias) in the analyzed studies. It encourages authors to make a clear distinction between the "quality of a study" and the "bias of a study" in the following way [18]:

Quality is often the best the authors have been able to do. For example, authors may report the results of surgical trials in which blinding of the outcome assessors was not part of the trial's conduct. Even though this may have been the best methodology the researchers were able to do, there are still theoretical grounds for believing that the study was susceptible to (risk of) bias.

We can argue that the quality of 3D scans of clothes is very much the result of the characteristics of the equipment used for the scan, the characteristics of the item that was scanned (for example, whether has a shiny surface, if it is unmovable) and the overall setup characteristics (such as light conditions). A failure to explain these characteristics in detail could be seen as introducing bias in the results. However, since our research question concerns the issue of how have scanners been used, rather than how successfully they have been used, we find that failing to properly document the characteristics of the setup is a problem of the quality of the study rather than bias in the study.

2.7. *Data Synthesis*

RO1 and RO2 were answered by means of quantitative statistical analysis of each variable taken into consideration (see Table 3). All studies were considered in the analysis.

RO3 has a more qualitative nature, aiming to assess the "quality of the study" according to the classification given by reviewers who are experts in the field. This quality analysis was undertaken according to four dimensions:

1. Hardware and software: How well was the hardware and software described?
2. Procedure: How well was the procedure described (e.g., light, scanner settings, and how to use the equipment)?
3. Objects: How well did the authors describe the objects that were scanned and their historical context?
4. Guidelines: Did the authors offer guidelines for people who would like to scan objects similar to those presented?

From this, our methodology for data synthesis was as follows:

1. Each of the selected studies were independently evaluated (reviewed) by Żyła, Kęsik and Santos in terms of these four dimensions.

2. For the dimensions of hardware and software, procedure, and objects, a Likert Scale was used with five categories: (1) not described, (2) partly described, (3) moderately described, (4) very well described, and (5) completely described.
3. The guidelines dimension was evaluated using a binary scale (1 = yes, 0 = no).
4. The assessments of the three reviewers were then aggregated by an average mean on each of the three first dimensions (hardware and software, procedure, and objects). We decided that the guidelines dimension should not be used in assessing the quality of the study, as this variable was mainly created to identify whether or not guidelines are frequently offered in the studies.
5. Finally, the three reviewers discussed any instances of very different item assessments between them. The final decision was reached by consensus.

The final outcome—an assessment of the general quality of each study—was reached by taking the average mean of the aggregated mean values of the three dimensions that were used to assess quality.

2.8. Comments on the PRISMA Protocol

The following elements of PRISMA protocol were not addressed: outcomes and prioritization, meta-bias(es), and confidence in cumulative evidence. We found them to be outside the scope of our article, due to the topic of study, the available data, and the character of the research objectives.

3. Results

The full list of studies that we found eligible for further analysis can be found in Table 4. We decided to assign a special ID that would be used to identify a study in the following sections using a simple code. Another code we introduced in the table concerns the type of study: CP—Conference Proceeding and JP—Journal Paper. In the following section, we analyze the studies in context of our research objectives.

Table 4. Full list of eligible studies.

No.	ID	Source	Title	Author	Study Type	Year
1	AC18	ACM Digital Library	Object-focused mixed reality storytelling: technology-driven content creation and dissemination for engaging user experiences	Darzentas et al. [21]	CP	2018
2	IE20	IEEE Xplore	Manipulating puppets in VR	Nitsche and McBride [22]	CP	2020
3	IE01	IEEE Xplore	Super high resolution 3D imaging and efficient visualization	Basu and Cheng [23]	CP	2001
4	SC14	Scopus	Reconstructing textile heritage	Calvert et al. [24]	JP	2014
5	SP18	Springer Link	3D visualization of a woman's folk costume	Kočevar et al. [11]	JP	2018
6	GS19a	Google Scholar	Future applications of digital clothing for historical costume: the past, present and future of fashion	McNulty [25]	CP	2019
7	GS19b	Google Scholar	Thickness mapping of body armor: a comparative study of eight breastplates from the National Museum of Slovenia	Lazar and Kraner [26]	JP	2019
8	GS18	Google Scholar	3D technology in collaborative heritage preservation	DeHass and Taitt [27]	JP	2018
9	GS17	Google Scholar	3D scanning and 3D printing technologies used in Albanian heritage preservation	Tota et al. [28]	JP	2017
10	GS16	Google Scholar	Use of image based modeling for documentation of intricately shaped objects	Marčiš et al. [29]	JP	2016
11	GS13	Google Scholar	Enhancing the display of the fashion artefact through digital multi-media approaches	Capacete-Caballero et al. [30]	CP	2013
12	GS12	Google Scholar	3D modeling of cultural objects in the V&A Museum: tools and workflow developments	Stevenson et al. [31]	JP	2012

3.1. Answer to Research Objective 1

To answer Research Objective 1 (RO1: Identify whether studies on this topic exist and how many there are) we conducted a bibliometric analysis on the selected studies. In general, a bibliometric analysis assumes that each field of study can be described by using significant information collected from the metadata of the literature published in that field, such as authors, keywords, terms from titles and abstracts, citations, and references [32].

We began by analyzing the frequency of published studies by year and by type (whether if it was published in a conference proceeding or a journal). Figures 2 and 3 show the results. In average one study per year was published starting from 2012. Only 1 of 12 studies was published before 2012—a technology (a laser scanner) showcase from 2001. Most studies (7 of 12) were published in journals, the rest in conference proceedings. Google Scholar was the source for seven works that were not indexed by the high-quality bibliographic databases.

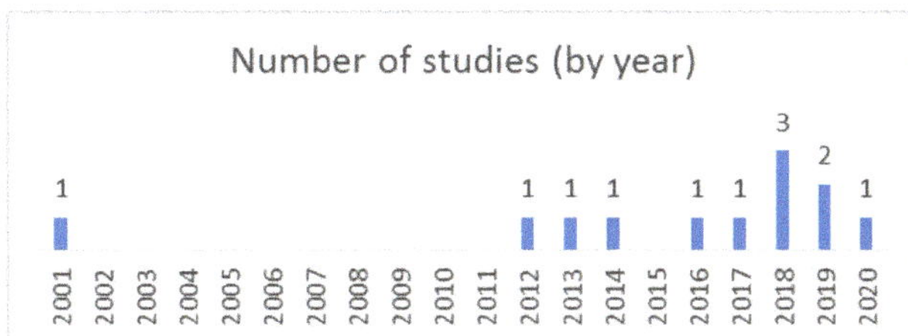

Figure 2. Number of studies by year.

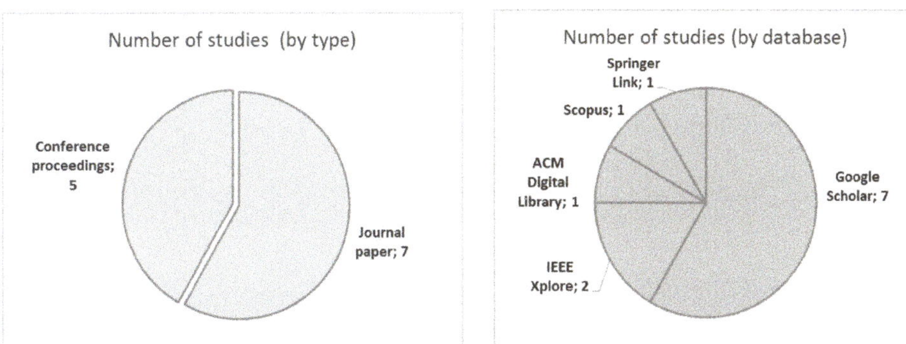

(**a**) (**b**)

Figure 3. Distribution of studies by (**a**) type and (**b**) database.

We then conducted an analysis on the keywords of each article to identify those used by the article's authors and also their frequency. We organized them into one of four groups:

1. Technology: terms related to the technology that was used, such as scanning, resolution, virtual reality, visualization.
2. Clothing: terms related to the item being scanned, such as textile, fabric, costume.
3. Museum: terms related to the museum's aim in digitizing the object, such as archiving, documentation, preservation.
4. Cultural: terms related to the historical or cultural context of the scanned item, such as Alaska native or Middle Ages.

To achieve this goal, some other procedures were carried out:

- "Compound" keywords were split; for example, the keyword "Gorenjska folk costume image analysis" was divided into three: "Gorenjska" (cultural), "folk costume" (clothing), and "image analysis" (technical);
- Keywords were aggregated by their similarity under a common simplified keyword when we found it pertinent. This happened with the keyword "National Museum of Slovenia", which was simplified to the more general (and widely used) keyword "museums".

Table 5 shows the keywords used, organized by type, and their frequency. The frequency of keywords is uniform—in average each keyword occurs once. The biggest group of keywords is technology-related (29 keywords), and the smallest one is culture-related (five keywords). Remaining groups are comparable in number, being 11 in the case of clothing, and 14 in the case of museum.

Table 5. Keywords by type and their frequency.

Keyword Group	Keywords
Technology	3D graphics (1); 3D imaging (1); 3D modeling (2); 3D printing (1); 3D scanning (3); 3D technology (1); 3D visualization (1); alpha map (1); augmented reality (1); digital (1); digital fashion (1); digital repatriation (1); efficient visualization (1); illustration (1); image analysis (1); image-based modeling (1); interaction design (1); interactive visualization (1); mixed reality (2); online participation (1); pattern design (1); prototype (1); storytelling (1); super high resolution (1); technology (1); thickness mapping (1); user engagement (1); user experience (1); virtual reality (2)
Clothing	accessories (1); armor (1); breastplate (1); folk costume (2); fragment (1); handicraft (1); puppetry (1); spiral staircase (1); textile (1); timber roof truss (1); woven fabric porosity (1)
Museum	archive (1); communication (1); conservation (1); cultural heritage (1); documentation (1); fashion curation (1); fashion film (1); historic preservation (1); historical fashion (1); historical furniture (1); innovation (1); museums (2); preservation (1); restoration (1)
Cultural	Alaska native (1); Early Modern period (1); Gorenjska (1); indigenous peoples (1); Middle Ages (1)

3.2. *Answer to Research Objective 2*

To answer Research Objective 2 (RO2: Identify the purpose driving the creation of 3D scans and the technology used) we conducted a qualitative analysis on the selected studies, extracting information about scanning devices, the type of technology used for 3D scanning, the purpose of the scan and the objects being scanned.

We classified the technology used into the following categories: structured light scanning, laser scanning, photogrammetry and computerized tomography scanning. We classified the purpose of scanning into six categories: dissemination, archiving (which might also be perceived as a form of documentation), reconstruction, technology presentation, research, and developing a digitization workflow. One study could be assigned up to two purposes in this categorization system, selecting the most important two where several were applicable. In some cases, we were not able to establish specific details; such instances are recorded as "not established" in Table 6.

The results of the qualitative analysis are shown according to the distribution of 3D technology type (Figure 4), the distribution of purposes for scanning (Figure 5) and the distribution of scanned object types (Figure 6). The most frequent technology choices were photogrammetry (six studies) and laser scanning (four studies). Other identified technologies occurred only once. The dominant purpose of 3D scanning was dissemination (eight studies) and archiving (four studies). In two studies, scans were made as a base for research. A majority (seven studies) concerned the scanning of full-size clothes; however, three studies focused on the scanning of historical clothed toys. The remaining publications concerned fragments of clothes (two studies) and armor (one study).

Table 6. Summary of technology and purpose of 3D scanning.

ID	Devices	Technology Type	Purpose	What was Scanned?
AC18	digital camera	photogrammetry	dissemination	personal objects (including clothes) brought by visitors of a museum
IE20	FaroArm	laser scanning	archiving, dissemination	puppets in clothes
IE01	TelePhotogenics Inc. Zoomage360	laser scanning	technology presentation	a doll in clothes
SC14	Nikon Metrology XTH 225 micro-CT scanner	computerized tomography scanning	archiving, reconstruction	a decorative collar
SP18 *	not established	not established	dissemination	a folk costume
GS13	device running Autodesk "123D Catch" app	photogrammetry	dissemination	historic garments
GS18	not established	photogrammetry	research, dissemination	ethnographic collections
GS17	Konica Minolta Vivid 910	laser scanning	archiving, dissemination	a waist coat
GS19a	device running Eyeexpo app	photogrammetry	dissemination	clothes collections
GS19b	not established	laser scanner and photogrammetry	research	armor, including a textile example
GS12	Breuckmann Smart Scan-HE	structured light scanning	developing digitization workflow	a small doll in clothes
GS16	Canon 5D MkIII and PhotoScan app	photogrammetry	archiving, dissemination	a folk costume

* 3D meshes were designed from scratch through Blender. Photographs and 3D scans were used to capture realistic and accurate images of the materials.

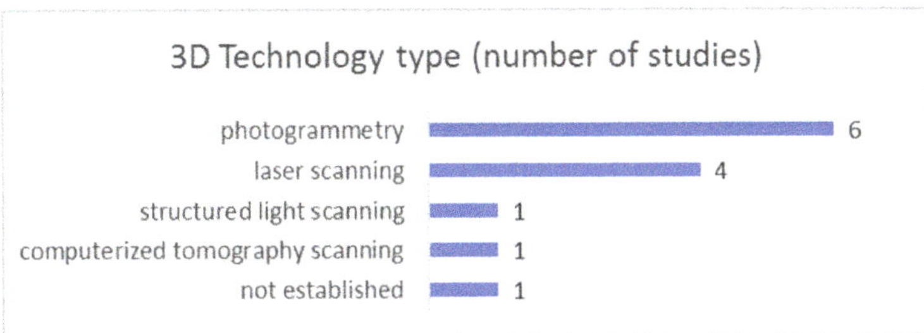

Figure 4. Distribution of 3D-technology types.

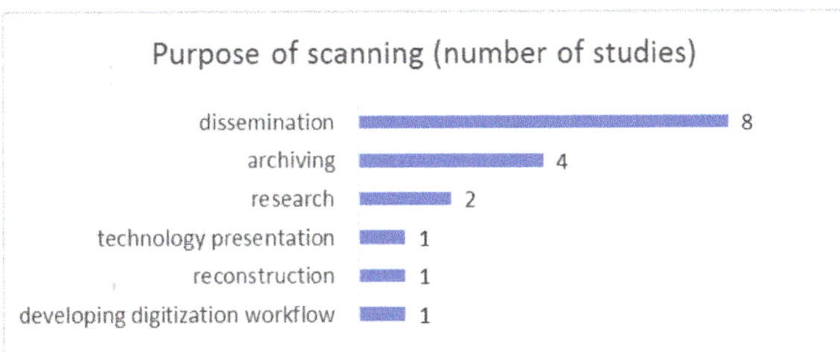

Figure 5. Distribution of reasons for scanning.

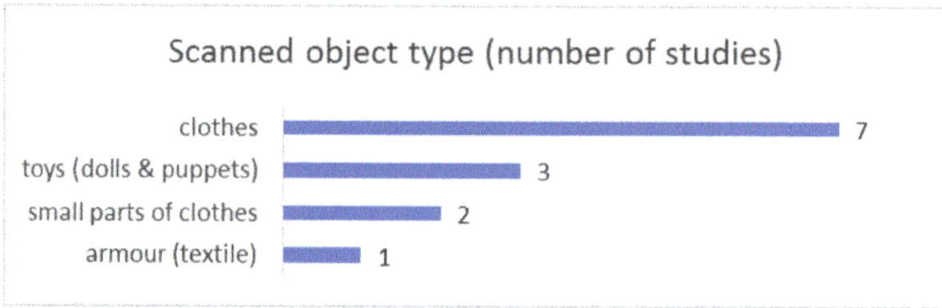

Figure 6. Distribution of a scanned object type.

3.3. Answer to Research Objective 3

To answer Research Objective 3 (RO3: Assess how well the literature describes the 3D scanning procedure), we conducted quantitative analysis on the selected papers. The five-point Likert scale was used to assess the general quality of the paper, as well as the quality of description of the hardware and software used for 3D scanning, the procedure of 3D scanning, and the background of the 3D-scanned objects. A Yes/No scale was used to determine whether the study was focused on providing guidelines. See Table 7 for details.

Table 7. Quality evaluation of identified studies.

ID	Hardware and Software	Procedure	Objects	Guidelines	General Quality
AC18	3	2	1	No	2
IE20	4	3	4	No	2
IE01	4	3	2	No	1
SC14	5	4	4	No	5
SP18	3	4	5	No	4
GS13	3	3	2	Yes	4
GS18	2	2	3	Yes	4
GS17	4	3	2	No	3
GS19a	2	1	2	No	2
GS19b	1	1	4	No	1
GS12	4	5	2	Yes	3
GS16	4	4	4	Yes	4
Average	**3**	**3**	**3**		**3**

The data seem to show a high degree of variation in the quality of the studies. Figure 7 shows the distribution graphs of the quality by each one of the three dimensions—hardware and software, procedure, and objects. The charts imply the following conclusions:

1. Hardware and software is the dimension the studies' authors describe with the greatest level of detail across the board, leading to a higher "quality" score: 50% of the studies received a score of 4 (good) or 5 (very good), while 75% of the studies were at least moderate (3).
2. The level of detail provided on procedure was also relatively good: at least 66% of the studies scored 3 (moderate) or above.
3. There appears to be greater heterogeneity in the quality of descriptions of objects: 50% of the studies were found to present a poor or very poor description of the objects.

An aggregation of the three dimensions of quality is presented in Figure 8. It illustrates that there is a great heterogeneity between the studies, having an almost uniform distribution between the five categories of quality (from very poor to very good).

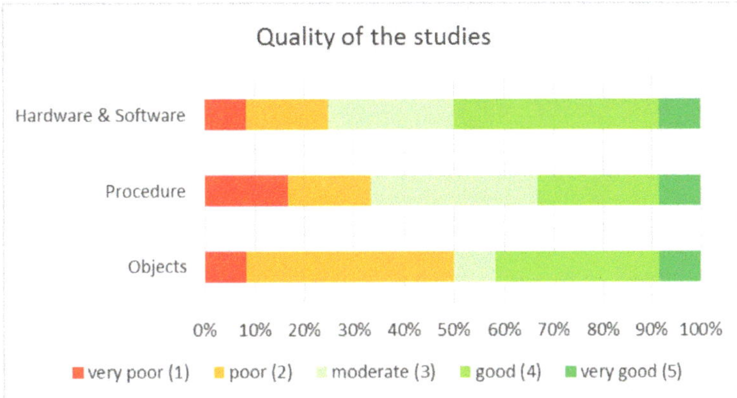

Figure 7. The quality of the selected studies.

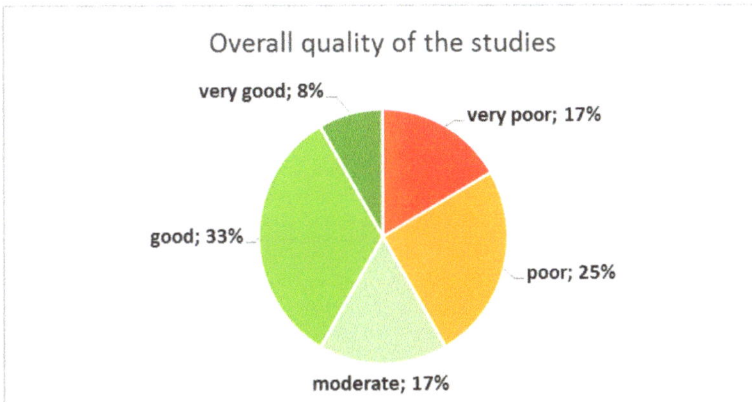

Figure 8. Overall quality of the selected studies.

4. Conclusions

This article discovers and compares methodologies of 3D scanning of historical clothes presented in the literature, in order to determine if a new methodology is needed. In order to answer the research question, How have 3D scanners been used to digitize historical clothes? we established three research objectives:

RO1: Identify whether studies on this topic exist and how many there are.

RO2: Identify the purpose driving the creation of 3D scans and the technology used.

RO3: Assess how well the literature describes the 3D scanning procedure.

In terms of RO1, after screening, we identified 12 publications on this topic, a mix of journal articles and conference proceedings. The first one dated from 2001, and the remaining 11 were all published between 2012 and 2020.

We can only speculate on the reasons why there are so few studies on the 3D scanning of historical clothes. One hypothesis, as stated in Section 1.1, is that the application of 3D scanning technology in the digitization of cultural heritage objects is still a relatively young field and knowledge about these techniques is not widespread. A second hypothesis is that authors simply do not feel it is relevant to publish on this specific type of object (historical clothes), perhaps considering the challenges of scanning these artefacts (such as the problem of shinny surfaces) to be the same as for other objects—in other words, that there is nothing new to be "discovered" or published. We would argue that this is not the case, as historical clothes have particular characteristics (e.g., size, texture, and shapable volume) that present specific challenges when it comes to scanning. Last, there is the hypothesis that museums still do not see 3D models as "relevant" for certain purposes. As the literature suggests, 2D photographs, 360° photographic views, 3D CAD models and photogrammetry-based 3D models are often preferred for purposes of dissemination.

Regarding RO2, to discover the reasons for 3D scanning, we found that a substantial majority of 3D-scanning projects were carried out for dissemination purposes (eight instances), followed by motivations linked to archiving (four) and research (two). This finding is unsurprising given the rising importance of digital images for facilitating public access to museum collections, whether online (for example, enabling people to explore collections without visiting the museum) or to be deployed in interactive digital media as part of an in-house exhibition.

In terms of the technologies used (also covered in RO2), it turned out that 40% of the articles were not, in fact, about true 3D scanning; in fact, they were on photogrammetry, which, according to the scientific literature, is a separate technology. We included them to highlight a degree of confusion among authors as to what does and does not constitute 3D scanning. After photogrammetry, the most common approach was laser scanning, which was discussed in four of the articles. The surprising finding here was the general under-utilization of structured light scanning, which in our opinion is potentially better-suited technology for 3D scanning of historical clothes, at least in comparison to the other scanning technologies identified.

Turning to RO3, we found very wide variability in the quality of articles, with 41% classed as good or very good, but 42% described as poor or very poor. Generally speaking, the articles tended to deal with technical issues, such as the hardware and software, quite thoroughly, followed by procedure. Fewer articles gave a thorough description of the objects themselves. However, ultimately, the question of whether the focus is on the technology, the procedure, or the object will depend on the specific topic of a given article and the aims of the research it describes.

Generally speaking, we found there was a lack of high-quality articles dealing with the 3D scanning of clothing, despite the fact that this technology is potentially suited to the digital preservation of these complicated amorphous objects. We also found that only a third of the articles offered guidelines about the process. This suggests there may be scope—or even a need—for further studies exploring the relevant techniques in more detail, thus helping to build a body of best practice among museum professionals and digitization specialists.

With this article, we aimed to systematize the state-of-the-art on this topic and; as we have seen, the field at this point remains a very small and fragmented one. We found that most of the identified studies failed to provide a fair and complete description of the methodology used. We believe that authors should thoroughly document all technical aspects of their 3D scanning process, including the hardware and software used, the procedures undertaken, and the conditions in which the scanning was performed (such as the room, the light setup, and so on). A detailed explanation of the scanned artefact should also be provided, covering its size and shape but also every aspect of the fabric that was taken into consideration (such as textures and colors). It is also important to explain the purpose of the digitization project (for example, dissemination or preservation), as we believe this greatly influences the procedure (for example, detail is more important for preservation than for dissemination).

Three-dimensional scanning technology has already proven to be of tremendous value for museums in their mission to preserve, study, and disseminate history and culture. The topic of the 3D scanning of historical clothes is clearly under-explored, especially in the perspective of UN and EU initiatives and funding. Our analysis revealed that a complex and coherent methodology proposal that could address these needs is missing, and this was a tremendous motivational trigger—as a result, such methodology was proposed in another article [33]. We hope that this article offers a small but important contribution to bringing this topic to light and allotting it the importance it deserves.

Author Contributions: Conceptualization, K.Ż., F.S., J.K.; methodology, F.S. and K.Ż.; software, K.Ż.; validation, G.H.; formal analysis, F.S. and K.Ż.; investigation, K.Ż., J.K. and F.S.; resources, J.K., K.Ż., G.H. and F.S.; data curation, K.Ż., J.K. and F.S.; writing—original draft preparation, F.S. and K.Ż.; writing—review and editing, K.Ż., F.S., G.H., J.K.; visualization, F.S.; supervision, K.Ż.; project administration, K.Ż.; funding acquisition, J.K. and K.Ż. All authors have read and agreed to the published version of the manuscript.

Funding: This article was financed by the Lublin University of Technology.

Institutional Review Board Statement: Not applicable.

Informed Consent Statement: Not applicable.

Data Availability Statement: Not applicable.

Acknowledgments: We would like to thank the Centre for Studies in Education and Innovation (CI&DEI) and the Polytechnic of Leiria for their support.

Conflicts of Interest: The authors declare no conflict of interest.

References

1. Żyła, K.; Montusiewicz, J.; Skulimowski, S.; Kayumov, R. VR technologies as an extension to the museum exhibition: A case study of the Silk Road museums in Samarkand. *Muzeológia Kultúrne Dedičstvo* **2020**, *8*, 73–93. [CrossRef]
2. Petrelli, D. Making virtual reconstructions part of the visit: An exploratory study. *Digit. Appl. Archaeol. Cult. Herit.* **2019**, *15*, 1–12. [CrossRef]
3. Thwaites, H.; Santano, D.; Esmaeili, H.; See, Z.S. A Malaysian cultural heritage digital compendium. *Digit. Appl. Archaeol. Cult. Herit.* **2019**, *15*, 1–8. [CrossRef]
4. Milosz, M.; Kęsik, J.; Montusiewicz, J. 3D scanning and visualization of large monuments of Timurid Architecture in Central Asia-A methodical approach. *J. Comput. Cult. Herit.* **2020**, *14*, 1–31. [CrossRef]
5. Kuttruff, J.T.; Strickland-Olsen, M. Handling archaeological textile remains in the field and laboratory. In *Beyond Cloth and Cordage: Archaeological Textile Research in the Americas*; Drooker, P.B., Webster, L.D., Eds.; University of Utah Press: Salt Lake City, UT, USA, 2002; pp. 25–50.
6. Szabóová, N. Úspechy výšiviek z produkcie Spolku Izabella na medzinárodnom trhu. *Muzeológia Kultúrne Dedičstvo* **2018**, *6*, 95–103.
7. Fashion Museum Bath. Dress of the Year .... Available online: https://www.fashionmuseum.co.uk/galleries/dress-year (accessed on 15 March 2021).
8. Cristóbal Balenciaga Museoa. Fashion and Heritage. Available online: https://www.cristobalbalenciagamuseoa.com/en/discover/digital-exhibitions/online-exhibitions/cristobal-balenciaga-fashion-and-heritage-.html (accessed on 15 March 2021).
9. Wijnhoven, M.A.; Moskvin, A. Digital replication and reconstruction of mail armour. *J. Cult. Herit.* **2020**, *45*, 221–233. [CrossRef]
10. Villarreal, N. The Use of 3D Apparel Simulation Software for Digitizing Historic Costume. Ph.D. Thesis, North Carolina State University, Raleigh, NC, USA, 2020.
11. Kočevar, T.N.; Nagič, B.; Gabrijelčič Tomc, H. 3D visualisation of a woman's folk costume. In *Digital Cultural Heritage. Lecture Notes in Computer Science*; Springer: Cham, Switzerland, 2018; Volume 10605, pp. 304–323. [CrossRef]
12. Ceccarelli, S.; Guarneri, M.; Ferri de Collibus, M.; Francucci, M.; Ciaffi, M.; Danielis, A. Laser scanners for high-quality 3D and IR imaging in cultural heritage monitoring and documentation. *J. Imaging* **2018**, *4*, 130. [CrossRef]
13. Santos, P.; Ritz, M.; Fuhrmann, C.; Fellner, D. 3D mass digitization: A milestone for archeological documentation. *Virtual Archaeol. Rev.* **2017**, *8*, 1–11. [CrossRef]
14. Farella, E.; Menna, F.; Nocerino, E.; Morabito, D.; Remondino, F.; Campi, M. Knowledge and valorization of historical sites through 3D documentation and modeling. *Int. Arch. Photogramm. Remote Sens. Spat. Inf. Sci.* **2016**, *41*, 255–262. [CrossRef]
15. Giannattasio, C.; Papa, L.M.; D'Agostino, P.; D'Auria, S. The BIM model for existing building heritage: From the geometric data acquisition to the information management. In *Graphical Heritage. EGA 2020. Springer Series in Design and Innovation*; Springer: Cham, Switzerland, 2020; Volume 5, pp. 311–322. [CrossRef]
16. Remondino, F. Geomatics and cultural heritage. In *3D Recording and Modelling in Archaeology and Cultural Heritage-Theory and Best Practices*; BAR Publishing: Oxford, UK, 2014; pp. 13–14.
17. Moher, D.; Shamseer, L.; Clarke, M.; Ghersi, D.; Liberati, A.; Petticrew, M.; Shekelle, P.; Stewart, L.; PRISMA-P Group. Preferred reporting items for systematic review and meta-analysis protocols (PRISMA-P) 2015 statement. *Syst. Rev.* **2015**, *4*, 1–9. [CrossRef] [PubMed]
18. Liberati, A.; Altman, D.G.; Tetzlaff, J.; Mulrow, C.; Gøtzsche, P.C.; Ioannidis, J.P.A.; Clarke, M.; Devereaux, P.J.; Kleijnen, J.; Moher, D. The PRISMA statement for reporting systematic reviews and meta-analyses of studies that evaluate health care interventions: Explanation and elaboration. *BMJ* **2009**. [CrossRef] [PubMed]
19. Radua, J. PRISMA 2020–An updated checklist for systematic reviews and meta-analyses. *Neurosci. Biobehav. Rev.* **2021**, *124*, 324–325. [CrossRef] [PubMed]

20. Shamseer, L.; Moher, D.; Clarke, M.; Ghersi, D.; Liberati, A.; Petticrew, M.; Shekelle, P.; Stewart, L.A.; the PRISMA-P Group. Preferred reporting items for systematic review and meta-analysis protocols (PRISMA-P) 2015: Elaboration and explanation. *BMJ* **2015**, 1–25. [CrossRef] [PubMed]
21. Darzentas, D.; Flintham, M.; Benford, S. Object-focused mixed reality storytelling: Technology-driven content creation and dissemination for engaging user experiences. In Proceedings of the 22nd Pan-Hellenic Conference on Informatics (PCI '18), Athens, Greece, 29 November–1 December 2018; ACM: New York, NY, USA, 2018; pp. 278–281. [CrossRef]
22. Nitsche, M.; McBride, P. Manipulating puppets in VR. In Proceedings of the 2020 IEEE Conference on Virtual Reality and 3D User Interfaces (VR), Atlanta, GA, USA, 22–26 March 2020; pp. 10–17. [CrossRef]
23. Basu, A.; Cheng, I. Super high resolution 3D imaging and efficient visualization. In Proceedings of the Joint 9th IFSA World Congress and 20th NAFIPS International Conference (Cat. No. 01TH8569), Vancouver, BC, Canada, 25–28 July 2001; pp. 346–351. [CrossRef]
24. Calvert, S.; Power, J.; Ryall, H.; Bills, P. Reconstructing textile heritage. *J. Writ. Creat. Pract.* **2014**, *7*, 415–425. [CrossRef]
25. McNulty, R. Future applications of digital clothing for historical costume: The past, present and future of fashion. In Proceedings of the EVA London 2019 (EVA 2019), London, UK, 8–11 July 2019; pp. 239–242. [CrossRef]
26. Lazar, T.; Kraner, J. Thickness mapping of body armour: A comparative study of eight breastplates from the National Museum of Slovenia. *Fasc. Archaeol. Hist.* **2019**, *32*, 129–145. [CrossRef]
27. DeHass, M.C.; Taitt, A. 3D technology in collaborative heritage preservation. *Mus. Anthropol. Rev.* **2018**, *12*, 120–153. [CrossRef]
28. Tota, A.; Shehi, E.; Onuzi, A. 3D scanning and 3D printing technologies used in Albanian heritage preservation. *Eur. J. Eng. Technol. Res.* **2017**, *2*, 39–45. [CrossRef]
29. Marčiš, M.; Barták, P.; Valaška, D.; Fraštia, M.; Trhan, O. Use of image based modelling for documentation of intricately shaped objects. *Int. Arch. Photogramm. Remote Sens. Spat. Inf. Sci.* **2016**, *XLI-B5*, 327–334. [CrossRef]
30. Capacete-Caballero, X.; Caulfield-Sriklad, D.; McKay, F. Enhancing the Display of the Fashion Artefact through Digital Multi-Media Approaches. In *Proceedings of the 1st International Conference on Digital Fashion*; Assn for Computing Machinery: London, UK, 2013; pp. 336–345. Available online: https://www.researchgate.net/profile/Giovanni-Maria-Conti/publication/274698158_Twyle_Play_with_fashion_intelligently/links/552563110cf24b822b40220c/Twyle-Play-with-fashion-intelligently.pdf (accessed on 16 June 2021).
31. Stevenson, J.; Jimenez, C.; Kelleher, P.; Knox, U. 3D modelling of cultural objects in the V&A Museum: Tools and workflow developments. In Proceedings of the IS&T Archiving Conference, Copenhagen, Denmark, 12–15 June 2012; pp. 168–173.
32. Li, J.; Antonenko, P.D.; Wang, J. Trends and issues in multimedia learning research in 1996–2016: A bibliometric analysis. *Educ. Res. Rev.* **2019**, *28*, 1–21. [CrossRef]
33. Montusiewicz, J.; Miłosz, M.; Kęsik, J.; Żyła, K. Structured-light 3D scanning of exhibited historical clothing-A first-ever methodical trial and its results. *Herit. Sci.* **2021**, *9*, 1–20. [CrossRef]

applied sciences

MDPI

Article

Methodology of 3D Scanning of Intangible Cultural Heritage—The Example of Lazgi Dance

Maria Skublewska-Paszkowska [1], Pawel Powroznik [1], Jakub Smolka [1], Marek Milosz [1,*], Edyta Lukasik [1], Dilbar Mukhamedova [2] and Elzbieta Milosz [1]

[1] Department of Computer Science, Lublin University of Technology, Nadbystrzycka 36B, 20-618 Lublin, Poland; maria.paszkowska@pollub.pl (M.S.-P.); p.powroznik@pollub.pl (P.P.); j.smolka@pollub.pl (J.S.); e.lukasik@pollub.pl (E.L.); e.milosz@pollub.pl (E.M.)

[2] Department of Psychology, National University of Uzbekistan, VUZ Gorodok, Tashkent 700174, Uzbekistan; sdilbar@yandex.ru

* Correspondence: m.milosz@pollub.pl

Abstract: Traditional dance is one of the key elements of Intangible Culture Heritage (ICH). Many scientific papers concern analysis of dance sequences, classification and recognition of movements, making ICH data public, creating and visualising 3D models or software solutions for learning folklore dances. These works make it possible to preserve this disappearing art. The aim of this article is to propose a methodology for scanning folklore dances. The methodology was developed on the basis of capturing 3D data via an optical motion capture system with a full body Plug-in Gait model that allows for kinematic and kinetic analysis of motion sequences. An additional element of this research was the development of a hand model with which it is possible to precisely analyse the fingers, which play a significant role in many dances. The present methodology was verified on the basis of the Lazgi dance, included in the UNESCO ICH list. The obtained results of movement biomechanics for the dance sequence and the angles of the fingers indicate that it is universal and can be applied to dances that involve the upper and lower body parts, including hand movements.

Keywords: dance scanning; dance capturing methodology; motion capture; Lazgi dance

Citation: Skublewska-Paszkowska, M.; Powroznik, P.; Smolka, J.; Milosz, M.; Lukasik, E.; Mukhamedova, D.; Milosz, E. Methodology of 3D Scanning of Intangible Cultural Heritage—The Example of Lazgi Dance. *Appl. Sci.* **2021**, *11*, 11568. https://doi.org/10.3390/app112311568

Academic Editor: João M. F. Rodrigues

Received: 7 November 2021
Accepted: 1 December 2021
Published: 6 December 2021

1. Introduction

According to UNESCO definition, performing arts, including dance, are an example of Intangible Culture Heritage (ICH), among such other activities as: oral traditions, expressions (e.g., language, storytelling), social practices, rituals, festive events and traditional crafts [1].

Folk dances were created over the centuries based on national tradition and ideology. They represent the dominant and distinctive musical and movement characteristics. Through the influence of the national culture, dance reflects awareness of national distinctiveness. In other words, dance maintains social cohesion along with cultural diversity [2]. In the modern world the art of folk dancing is disappearing. The style of dance passing from generation to generation is fragile, even haphazard [3]. That is why the preservation of folk dances plays such an important role in ICH. The accurate recognition of these moves is a challenging task.

The dynamic development of three-dimensional (3D) technology allows both to capture and protect this type of threatened heritage. 3D dance scanning is the process of capturing real-world dance in order to create a 3D model reflecting its shape and moves. Nowadays there are many sophisticated devices that allow dance acquisition, e.g., motion capture systems. The optical ones based on retro-reflecting markers attached directly to the human body or to a special suit allow capturing a dancer's silhouette. A similar way of movement acquisition can be done with the use of active-markers based on LED technology that enable capturing movements indoors and outdoors. The capturing of dancers may also be obtained using various sensors, such as Kinect or Xsens. The 3D data collected is further used for detailed analysis or displaying the moves in various software (e.g., for developing Virtual Reality). This process is also gaining further understanding from ICH.

Lazgi is an ancient dance created by the inhabitants of the downstream areas of the Amu-Daria River. Its history is closely related to the ancient land of Khorezm, which covers the area south of the Aral Sea. There is an opinion that the ancient meaning of the word "lazgi" is trembling, characterised by vigorous body movement and facial expression [4], which very vividly reflects the characteristics of the dance. There are many legends and myths associated with the rise of the dance [4]. The most widespread is the legend of the soul-body connection thanks to the rhythms of Lazgi. After the act of creating the human body, it could not be brought to life because of the soul's reluctance to enter the body. The soul's attitude changed when it heard the melody of the Lazgi dance and began to slowly penetrate the body. Therefore, Lazgi starts with a leisurely introduction, then the dancer takes a position with a hand raised to the sun. Next begins a sequence of movements involving the fingers, wrists, arms, torso and legs. The whole body starts to move by a combination of turns and swings and the rhythm of the dance gradually speeds up [5]. Despite the emergence of numerous wars, natural changes, disasters and social unrest, the Lazgi dance has survived for thousands of years. In 2019 it was inscribed on the Representative List of the Intangible Cultural Heritage of Humanity UNESCO [6].

The aim of the paper is to develop a universal dance scanning method using the motion capture system with the biomechanical Plug-in Model. The retro-reflecting markers are attached to a special suit according to this model in order to capture the dancer's movements. Additionally, the markers were attached to the dancer's hands in order to capture hand gesticulation. The created hand model is dedicated for capturing a type of dance where the analysis of hand and finger movements is essential. The proposed methodology consists of the following steps: planning the research, motion capture session preparation and implementation and, finally, data post-processing. The analysis of the dance, which verified the proposed methodology, involved indicating the trajectories for the selected parts of the upper and lower body and fingers, as well as changes in angles.

2. Traditional Dance Studies—Literature Overview

There are many traditional dances in the literature study that have been analysed in various aspects of ICH research. The detailed results were gathered in Table 1.

Table 1. Traditional dances in the selected countries.

Country	Dance
Austria	Krebspolka [7]
Belgium	Walloon [8]
China	Lion Dance [9]
	Agogo [10]
	Dai, Uygur [11]
Cyprus	Zeimpekiko [12]
Czech Republic	Pašovská Sedlcká [13]
Greece	Laisios or Lagisios dance at a Greek Thracian wedding, Gikna [2],
	Syrtos [14–23],
	Syrtos Makedonikos [17,21–23]
	Syrtos Kalamatianos [14,15,17,19–22],
	Trehatos [14,15,19–21]
	Enteka [19–22]
	Tsamiko [24–26]
	Zeibekiko [27]
	Antikristos [27]
Hungary	Kalocsai mars [28]

Table 1. *Cont.*

Country	Dance
India	Bharatanatyam [29] Kathak [29] Kuchipudi [29] Odissi [29] Kathakali [29] Sattriya [29] Manipuri [29] Mohiniyattam [29]
Indonesia	Bedhaya Ketawang [30]
Japan	Musume-doujouji [31]
Korea	Didim [32]
Macedonia	Trexatos [22]
Portugal	Joao Fiadeiro's choreographic [33]
Slovakia	Horehronie, Abov, Podpol'anie and Horné, Považie [34]
Thailand	[35–38]
United Arab Emirates	Ayala [39]

The most popular methods for 3D capturing are divided into two main categories: active methods (laser scanners, range finders, structured light projectors) and passive methods (stereo vision and visual hulls) [22]. The technologies for traditional dance capturing are presented in Table 2.

Table 2. Technologies used to capture traditional dance data.

Data Collection Methods	Papers
3D visualization, 3D poses estimation from a video	[2]
Vicon motion capture system	[10,14–18,21,23,31]
OptiTrack motion capture system	[7,13,28,34]
Phasespace4 Impulse X2 motion capture system	[12,27]
Microsoft Kinect	[8,19–22,24–26,30,33]
Gyroscopes and accelerometers	[32]
Smartphones	[9]
Video camera	[11,29,33]
Labanotation system	[35–38]
IMU sensors	[39]

Five main aspects can be distinguished in the research: analysis of selected dance sequences, classification and pattern recognition, visualization and software solutions, description and verification of methodology and creation databases.

2.1. *Analysis of Selected Dance Sequences*

The first type of research concerns the analysis of trajectories obtained from motion capture systems, which show the dance rhythm and body kinematics and kinetics parameters. It allows for a better understanding of performance of the dance [28].

In [14] the studies concerning choreographic sequences analysis were performed based on 3D motion capture data. They involved trajectories of hip and foot joints showing the rhythm and periodicity of Greek dance. The average Silhouette value over all frames were also indicated for the particular dancers. In [16] the analysis of hip, knee and ankle joint angles was performed. Male dancers had narrower angles in all the mentioned joints of all the six steps than female dancers during the implementation of the Greek Syrtos dance. Additionally, in [15], the dance sequences were analysed in terms of average precision and recall values for different frequency. The identification of the Syrtos dance based on rhythm is also presented in [17]. The paper in [39] highlights a scientific approach to capture and analyse the biomechanical data of authentic traditional dance motions with the aim to digitally preserve their disappearing inherent rhythmic dying features for future training and coaching purposes. Inertial Measurement Unit (IMU)-based sensors were used to capture three-dimensional movement patterns of a professional traditional Ayala dancer. In [18], the choreography of the same Greek dance was analysed. Optimal key points were distinguished from 3D trajectories that best represented the choreography. These few frames are suitable for modelling an entire choreography, which significantly reduced the amount of information required for processing and storage. It is a particularly important aspect in modelling, analysing and storing this type of data. A Bayesian optimised, bi-directional Long Short-Term Memory (LSTM) network for pose characterisation of a choreography for Greek dance was proposed in [20]. In [21], the identification of dance choreographies based on skeleton data points was used for data obtained from heterogeneous motion capture systems (Vicon motion capture system and Kinect sensor). Finger movements based on gyroscopes and accelerometers attached to the joints for Korean dance were studied in [32]. An example analysis of traditional Hungarian dance analysis considering lower body kinematics, kinetics, biomechanical parameters (Centre of Pressure) and balancing ability through dance movements was described in [28]. The kinematic parameters, such as: Range of Motion (ROM) of knee flexion angle, ROM of hip flexion angle, and ROM of pelvis tilt angle were taken into the consideration. The hip and knee flexion-extension kinetic parameters were calculated for the successive dance steps. In [23], a deep stacked auto-encoder scheme was followed by a Hierarchical Sparse Modelling for Representative Selection (H-SMRS) summarisation algorithm for performing accurate synopses of three Greek dance sequences. Two dancers performed the following dances simultaneously. The analysis concerned human joint points from 3D data. In [25], the Tsamiko dance motion patterns from skeletal animation data captured by multiple Kinect sensors were performed. The skeletal data were split into five different body parts (torso, left hand, right hand, left leg and right leg), which were then transformed to allow invariant posture recognition.

2.2. Classification and Pattern Recognition

The second type of research concerns the classification and pattern recognition aspects. A clustering-based method for the selection of the basic primitives of a choreography, and a kinematics-based method that generates meaningful summaries at hierarchical levels of granularity were applied. It was also stated that the sex of the dancer may be recognised using 3D trajectories of toe markers and the kinematic data of joint angles [16]. The assessment of the Greek dances choreography at different representational levels was presented in [19]. The coarse levels showed the main steps of a Greek dance while the detailed levels provided an assessment per dance frame. The whole framework was created on the basis of artificial intelligence and pose identification (including bidirectional LSTM model with 3D kinematic data like joint velocity and acceleration). The effective classification and recognition of eight traditional Indian dances (accuracy score of 0.911) was presented in [29]. Deep Convolutional Neural Networks (DCNN), using ResNet50 as the base model by fine-tuning the last 14 layers was applied. Detection of basic traditional Walloon dance patterns was described in article [8]. The use of hidden Markov models to identify such steps as: Maclotte Base, Passepied Base, Passepied Fleuret and Back Step were characterised. Kinect was used as a movement registration tool. Three-dimensional position of 68 markers corresponding to 20 joints of the human body was used for classification. The classification efficiency was over 95%. As a result, it was decided to create a game that allows to learn Walloon dance. In [10], the use of motion capture technology and motion analysis method for learning traditional Chinese Agogo dance was presented. The system consists of 3D graphics, motion matching, motion database, and motion capture system. The user's movements, which were obtained by the motion capture system, were compared with the motions in the motion database through the motion-matching component. Additionally a 3D graphics component visualises the movements by the virtual teacher and user. To start learning, it was necessary to watch a 3D animation demonstrating the basic steps by professional dancers. Then the movement was captured in real time and rendered with a virtual representative, which was displayed next in a form of cylinders representing the body segments. The colour of a cylinder showed whether the position of the body segment was correct or not. In [11] the problem of dance video recommendation by exploring intrinsic motion components in dance videos as well as appearance information were investigated. The authors present the framework for a dance video recommendation method which consisted of several items: natural dance database, convolutional networks, random forests and more. The method was tested using the HIT Dance database containing 6 different types of dance videos, including Ballet, Hip-Hop, Chinese Classical Dance, Chinese Dai Folk Dance, Chinese Mongolian Folk Dance and Chinese Uygur Folk Dance.

2.3. Visualisation and Software Solutions

Visualisation and software solutions, such as Virtual Reality (VR), Augmented Reality (AR) and software platforms are the third type of research.

Visualisation in the form of application of Slovak folk dances recorded by using motion capture was presented in [34]. The effectiveness of learning steps from recorded dances as well as testing the overall usability of the application were verified. The game for teaching young generations of Cypriot folk dances as a way of promoting and preserving ICH was described in [12]. It was created based on 3D motion capture data. The motion of the user was captured in real time via Kinect. The game included different Skeletal Structure, body proportions and comparing motions. Another approach to the transmission of the traditional Greek Tsamiko dance in a form of game-based learning application was described in [24]. It consisted of several exercises, aiming to teach different variations of that dance. In [7] Krebspolka dance was recorded and further 3D models were used to create a mobile AR application. It provided a new way of learning and facilitate the process of learning a folk dance. This process was verified by motion capture sessions. The effect of transcranial Direct Current Stimulation (tDCS) on motor learning was analysed. It was stated that students with received tDCS perform dancing significantly better. In [13] a mobile AR application for assisting the process of learning folk dances based on 3D data was presented. Avatar representations (of either male or female) were synchronised with the digital representation of the dance. A great advantage of this system was the possibility of simultaneously learning to dance by males and females, because the dance was performed in pairs, male participants had to learn the male part of the dance and a male avatar was shown to them during the learning phase, while female participants had to learn the female version by observing a female avatar. In [27], the authors compared the movements of two avatars taking into consideration not only posture matching (meaning the physical geometry of the avatar) but also style, including the required effort, shape and interaction of the performer with the environment. Avatars were created based on 3D motion data. The presented algorithm focused on certain body parts, like upper/lower body and left/right side, as well as the whole body. Chinese performances of the Taiko drumming and lion dance recorded by smartphone were presented in [9]. This tool proved to be appropriate to register their sound. In [2] the CHROMATA platform is presented. It enabled the safeguarding of traditional dances as a part of ICH. It included analysis for 3D pose estimation, folklore dance recognition and textual analysis. Additionally, in [26], the Tsamiko dance was the topic of research where a novel methodology for dance learning and evaluation using multi-sensor and 3D gaming technology was described. Dance practitioner movements were continuously visualised with an avatar thanks to fused input from multiple sensors. The proposed system allowed to compare the quality of the student's movements with those of an expert due to the use of motion analysis and two-level Fuzzy Inference System. In [31] a method of acquiring torso movements during the traditional Japanese dance Musume-doujouji was presented. Previous studies focused primarily on the movements of the head, hands and feet, marginalising the movement of other parts of the body. The results of research using motion capture technology showed significant differences in the way of movements of the torso while recording the movements of professionals and amateurs constituted the basis of learning for beginners.

2.4. Description and Verification of Methodology

The fourth type of research concerns the methodology of dance registering.

The methodology of capturing Joao Fiadeiro's choreographic based on three Kinects and three video cameras was presented in [33]. The sequences for dance were also obtained. This method allowed for visualising the data by manipulating colour, time, and spatial position. The Labanotation system used to recode the gesture of Thai dance was presented in [35–38]. This tool was used to translate the notation scores of Thai dance movements into 3D animation. The methodology including choreographic analysis, 3D data capturing, 3D modelling of static and moving objects and symbolic representations for various traditional dances for the Terpsichore project was described in [22].

2.5. Creation Databases

The final type of research involved databases. In [30] the acquisition and processing methods of Bedhaya Ketawang—the traditional Indonesian dance—was presented. The authors proposed building knowledge for gesture 3D Modelling in Javanese Dance using a motion capture technique. As a result, a database of 3D model skeletons and gestures for building an interactive learning media of traditional Javanese dance was made. To achieve the research goal, a system consisting of a Kinect sensor and a set of animation tools was constructed. A very interesting concept of a database of dance movement sequences and their multi-faceted analysis is presented in [40]. The Dance Motion Capture Database (DMCD) discussed is defined by metadata in five main categories: descriptive, structural, administrative, reference and statistical. The DMCD contains text data about dance and multimedia. The text data describing the dance includes [40]: dance description, story-telling, country/region of origin, date when the dance first became known, type of the dance (e.g., solo or group) and its history. The multimedia collection primarily contains multimedia data of individual dance performers, such as the motion capture data, the video recordings and music. In addition to the multimedia records of individual dance performances, the DMCD stores the artist's data, the place where the records are made and the props used (i.e., clothes, additional items, etc.) The DMCD is used for the comparative analysis of dances and the search for similar sequences of movements in different dances. For this purpose, the concept of the Bag of Words was used [40]. This concept is widely used in machine learning, language processing and computer vision. This method is based on the representation of the text in the form of a multiset (bag) of words. In computer vision, the Bag-of-Visual-Words model is used to image classification [41]. By analogy, in [40], the Bag-of-Motifs (BoMs) motion model is used to represent the sequence of motion in dance for movements classification and searching for similar sequences. The 16 most informative joints with their relative joint angles analysed in 0.66 s long time windows were gathered to define motion words in [40]. The motion words were then grouped into clusters representing motion motifs and creating BoMs for dance. The database describing the dances can be used for further analysis of the dances.

3. Methodology of 3D Scanning of Dances

Dance is a complex combination of many different elements. It consists of choreography, music, clothes and other artefacts, scenography and others. Choreography is a sequence of dance movements performed by dancers ordered over time. It is characteristic of a given dance. The methodology of 3D dance scanning presents in the paper is limited to 3D digitisation of choreography, i.e., the movements of the dancer's body. One of the technologies that could be used is the Motion Capture (MC) technique with the use of optical systems. The developed methodology of 3D scanning of a dancer's movements and post-processing of the obtained results is presented in Table 3.

Table 3. Methodology of 3D digitisation of dance choreography.

Name of the Stage	Sub-Stages	Results
1. Planning	1.1. Obtaining knowledge about the dance 1.2. Selecting a music track 1.3. Technical planning MC session scheduling	Detailed plan of activities
2. MC session preparation	2.1. Acquiring dancers 2.2. Preparation of props 2.3. Booking a laboratory	Everything is ready for implementation of the MC session
3. MC session implementation	3.1. Participant preparation 3.2. Implementation of the session—data recording 3.3. Preliminary checking of the quality of the obtained data 3.4. Implementation of corrective recordings (if any) 3.5. Video recording of the dance	Raw data
4. Data post-processing	4.1. Post-processing and data cleaning 4.2. Creating a 3D choreography model 4.3. Indicating dance sequences 4.4. Analysis of the dancer's movements	3D model of the dance Sequence of elementary movements

The aim of the first stage of the work (Table 3) is to develop a detailed action plan. For this purpose, it is necessary to acquire knowledge about the dance and background music used in its implementation. Then it is necessary to plan activities in terms of the potential availability of the laboratory, equipment and technical staff. The work also includes detailed planning of the MC session. During the second stage of the methodology, the MC sessions are prepared. A dancer is commissioned, props are prepared (suitably sized costume, markers, etc.) A laboratory and technical staff are also booked during this stage. The MC session is carried out as the third stage of the methodology. During it, after the dancer is prepared (wearing a costume, placing markers in the right places), a session of recording 3D data is carried out. Videos from reference cameras are recorded simultaneously. During a break in the MC of the session, the quality of the acquired data is pre-checked and, if necessary, the selected dance sequences are repeated. After acquiring the 3D data, the dancer changes the costume to the national costume appropriate for the dance, and a video of the dance is recorded. During this stage, the right music is used, selected in the first stage. After collecting raw data, the stage of their post-processing follows—the fourth stage of the methodology (Table 3). Raw data are cleaned and processed (mainly in order to properly maintain the trajectories of the markers covered by the dancer). A 3D choreography model is then created from multiple movement sequences. This model is used for further analysis in order to search for elementary movement sequences.

4. Materials and Methods

4.1. Motion Capture System

In this study an optical motion capture Vicon Nexus system (Oxford Metrics Ltd., Oxford, UK) was applied. It consisted of eight T40S cameras operating in the near infrared, two Bonita reference video cameras, a Giganet hub collecting data and a desktop computer. The T40S cameras were mounted two on each wall, so that the central part of the room was clearly visible. The system records the movement of the markers placed on the special suit of the dancer. In order to register 3D marker position, it must be seen by at least two cameras. The reference cameras record video, which is used both for data processing and to generate video files containing integrated video and a biomechanical model of a dancer. A great advantage of the presented systems is the temporal integration of all the recorded data, both analogue and three-dimensional. The equipment is supplied with Vicon's Nexus software, 2.0 used to calibrate the system, record data and data processing.

4.2. Plug-in Gait Model

The motion capture system was integrated with the full body Plug-in Gait model [42], which determines the precise placement of 39 markers all over the human body. Its placement on the participant's body is presented in Figure 1 on the left. The model uses both upper and lower body for movement analysis. The model outputs calculate kinematic and kinetic data such as: joint angles, forces, moments and powers. The first three are given in three anatomical planes: sagittal, frontal and coronal. The outputs are calculated both for the right and left side. This model gives the possibility to calculate the biomechanics of the captured dance moves [43].

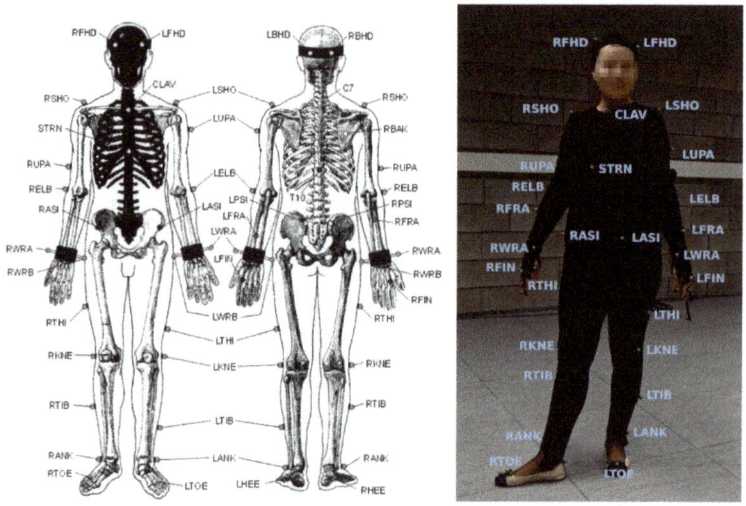

Figure 1. Full body Plug-in Gait marker placement [42] (on the **left**) and its real implementation (on the **right**).

4.3. Hand Model

Hand movement plays an important role in various dances [32,35]. The Plug-in Gait model does not contain markers on hands. Therefore, for the purposes of this research, a hand model was created based on the model [44]. It is presented in Figure 2. It consisted of 13 markers attached to the joints and tips of the fingers: two on the thumb (marked as TR), three on the pinky finger (marked as PR) and three on other fingers (marked as FR) except the marker (LFIN) which belongs to the Plug-in Gait model. Marker names started with L for the left hand and R for the right hand.

4.4. Dancer Preparation

The dancer wore a special suit that consisted of a sweatshirt, trousers and a cap. It was made of Velcro material, so that 14 mm retro-reflexive markers could be attached to it according to the Plug-in Gait model (Figure 1 on the right). In order to properly use this model, the participant had to be measured. The following dancer data had to be input to Vicon Nexus: height, weight, left and right leg length, right and left shoulder offset, left and right knee width, left and right ankle width, left and right elbow width, left and right wrist width and finally left and right hand width. They are all needed for calibration process and the proper model output calculations.

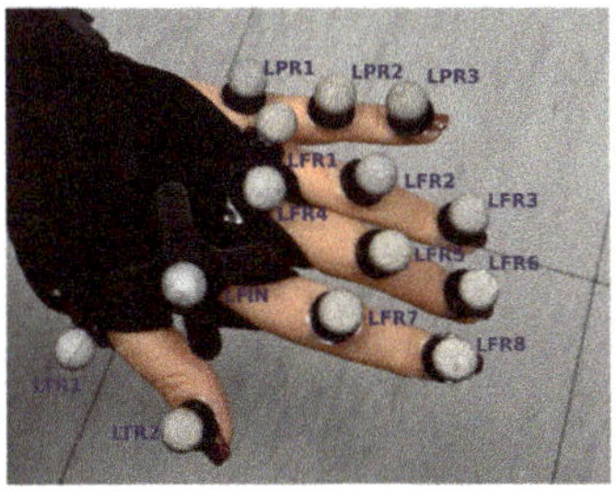

Figure 2. Left hand markers placement.

In order to perform the calibration, the static pose of the participant should be recorded. After indicating the labels for individual markers, all markers were verified if they were correctly attached. Calibration is the process which generates the skeleton for the captured person. Thanks to this, it will be possible to calculate outputs of the biomechanical model used.

4.5. Movement Capturing

Registration of movement was carried out with the T40S camera's resolution set to 100 Hz. The proper connection of all the components of the system was ensured by the use of the Nexus software. A series of proper tests was preceded by calibrating the system and setting the data management hierarchy. Movement acquisition was carried out automatically by the system and consisted in recording the subject's motion from start to finish (the moment of starting the recording depends on the operator). Movement was recorded during the playback of the Lazgi dance music. All the obtained recordings were linked to a specific person (subject). An important factor influencing the quality of the recording was the elimination of all reflective elements from the cameras' sight. Accidental glare resulting from elements of the subject's or the experimenter's outfit could introduce additional artefacts in the recordings and make subsequent processing more difficult. In this study the Lazgi dance was registered.

4.6. Data Post-Processing

Captured 3D data needed post-processing, which consisted of four main steps: labelling all markers, filling-in gaps, cleaning the recording and calculating the outputs for the biomechanical model. The first step assigned the proper label to a particular marker. It might be performed automatically or manually. The next step was to calculate the interpolated positions of markers based on the previous and subsequent ones. The continuous trajectory was obtained. The third step cleared the recording of additional unintended markers, such as instances of glare. Finally, for data prepared in this way, the outputs from the Plug-in Gait model were calculated. The obtained 3D data were ready for further analysis. The post-processed model of a dancer with the Plug-in Gait model applied, together with a hand model, is presented in Figure 3.

Due to the length of the recording of the entire dance, which took almost 4 min, post-processing was very time-consuming. A better solution would be to capture shorter dance sequences, or to divide the dance into sequences immediately after capturing and then perform the post-processing of the successive recordings.

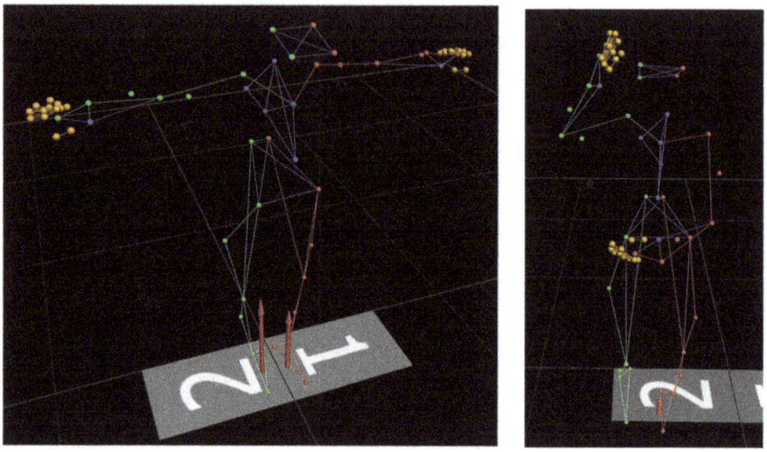

Figure 3. Model of the dancer after post-processing T-pose (on the **left**) and while performing the dance (on the **right**).

5. Motion Analysis

The methodology presented in this paper made it possible to conduct research, the results of which are both 3D silhouette data, 3D detailed hand data and angles for selected parts of body. The silhouette data relate to the position of 3D markers attached to the dancer's body (39 according to Plug-in Gait). The hand data correspond to 3D positions of 13 markers attached according the defined hand model. From the recorded Lazgi dance data, several sequences were obtained for various type of movements. For the purpose of motion analysis two dance sequences were selected for detailed analysis, including body and hand movements.

In the first sequence, the dancer bowed. The second sequence involved dynamic hand movements.

5.1. Body Movements

The analysis of the dancer's body movements concerns both the upper and lower parts. The obtained results include trajectories for selected markers, as well as kinematic analysis showing changes in angles for individual body parts. Trajectories presented were prepared in two planes: XY (transverse) and YZ (sagittal). The trajectories for upper body parts including shoulders (right and left) and wrists (right and left) from the thumb side indicted by the attached markers are presented in Figure 4.

The trajectories for lower body parts concerning hips (right and left) and knees (right and left) are presented in Figure 5.

For the second dance sequence the trajectories for the upper body parts including shoulders (right and left) and elbows (right and left) are presented in Figure 6.

For the second dance sequence the trajectories for the lower body parts concerning hips (right and left) and knees (right and left) are presented in Figure 7.

The angles for the lower body parts concerning ankles (right and left), knees (right and left) and pelvis (right and left) are presented in Figure 8. Please note that the lines for pelvis angles in Figure 8 are identical.

The angles for upper body parts concerning wrists (right and left), elbows (right and left) and shoulders (right and left) are presented in Figure 9.

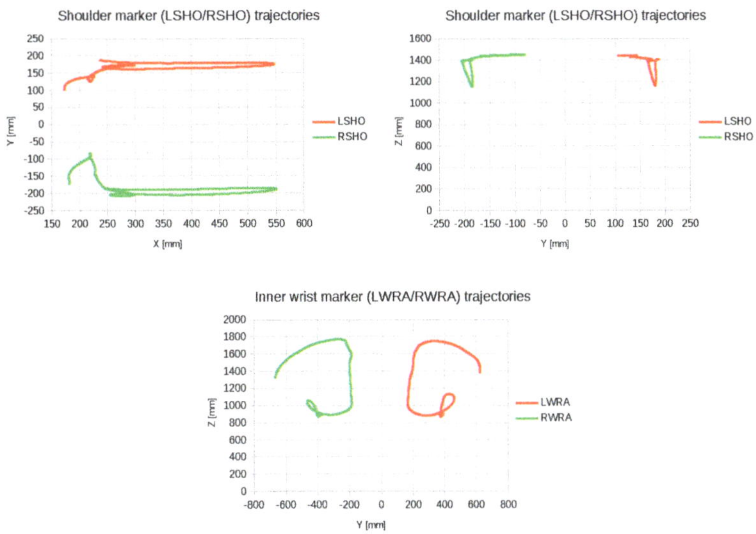

Figure 4. Trajectories for upper body parts for the first sequence.

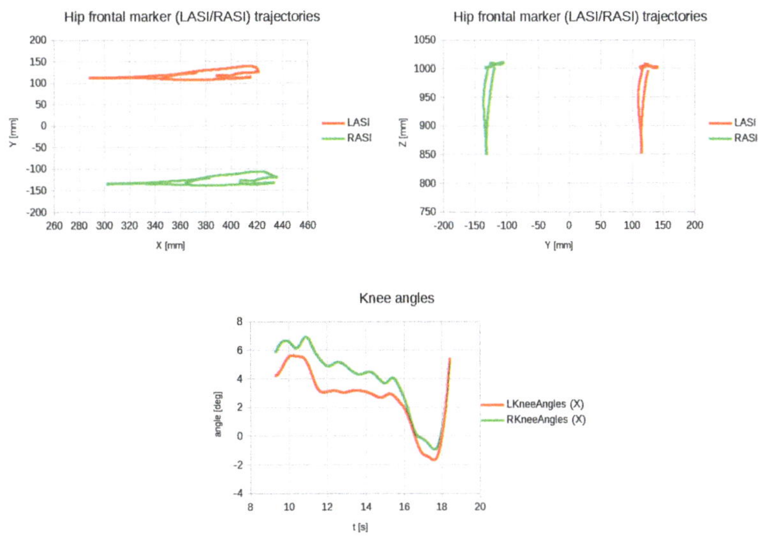

Figure 5. Trajectories for lower body parts for the first sequence.

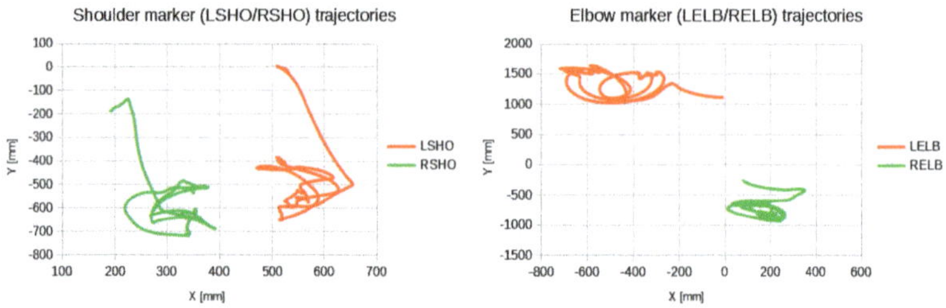

Figure 6. Trajectories for upper body parts for the second sequence.

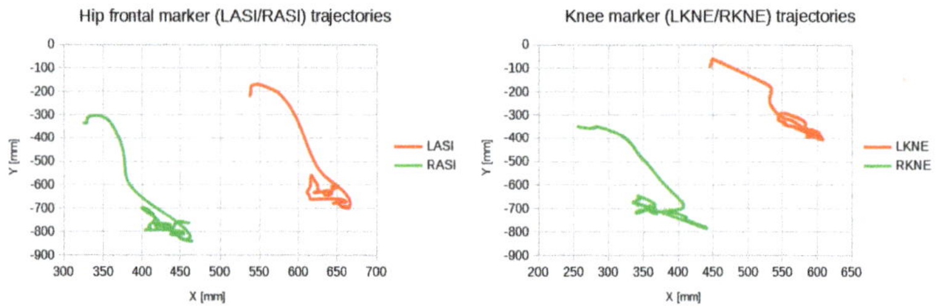

Figure 7. Trajectories for lower body parts for the second sequence.

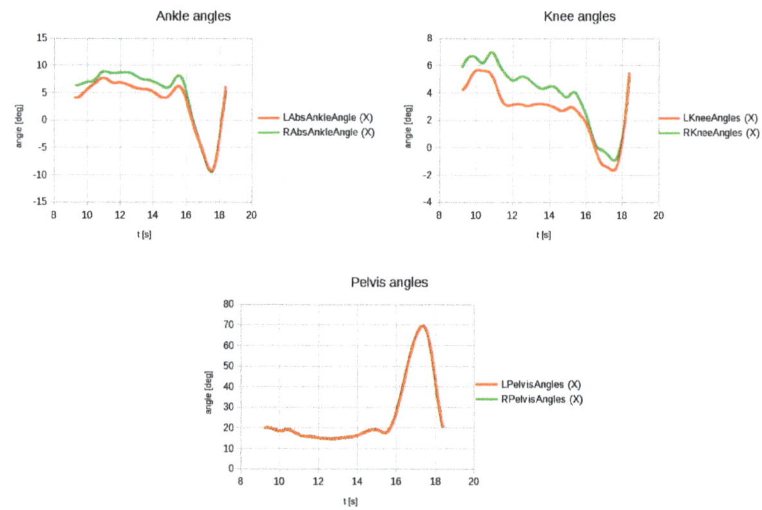

Figure 8. Angles for lower body parts for the first sequence.

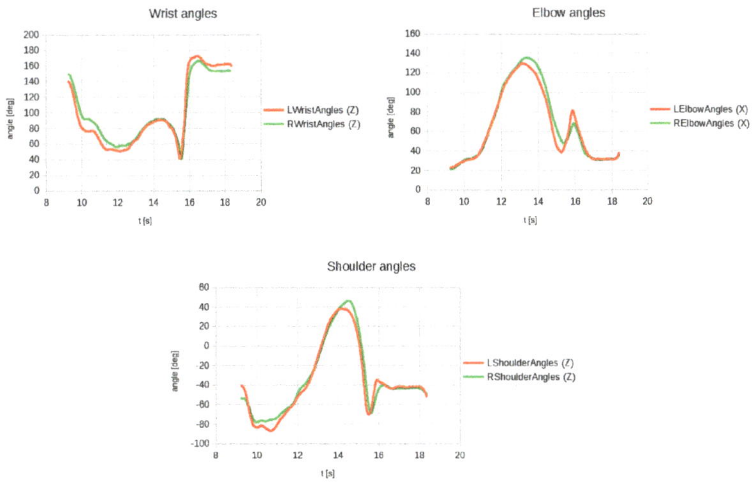

Figure 9. Angles for upper body parts for the first sequence.

5.2. Hand Movements

A dynamic dance sequence was selected for the analysis of hand movements (for four fingers, without thumb). The trajectories for the markers (LFR8 and RFR8) from the hand model are shown for the left and right hand, respectively. Two planes were analysed: XY and YZ. The results are presented in Figure 10.

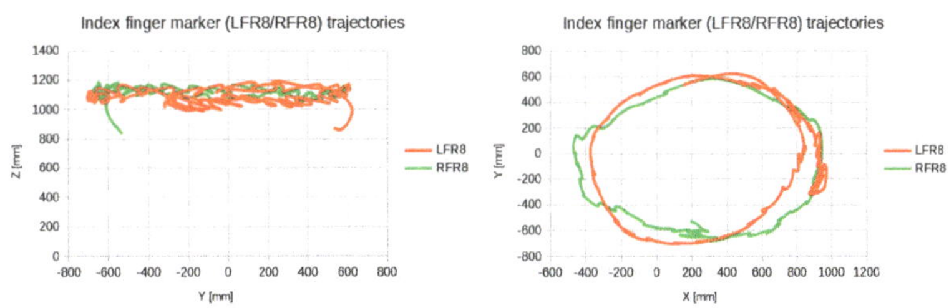

Figure 10. Trajectories for fingers for the second sequence.

The bend angles of the four fingers (ring, middle, index and little finger) were also calculated for each hand. The angles were calculated between the middle phalanges and the proximal phalanges. The results are presented in Figure 11.

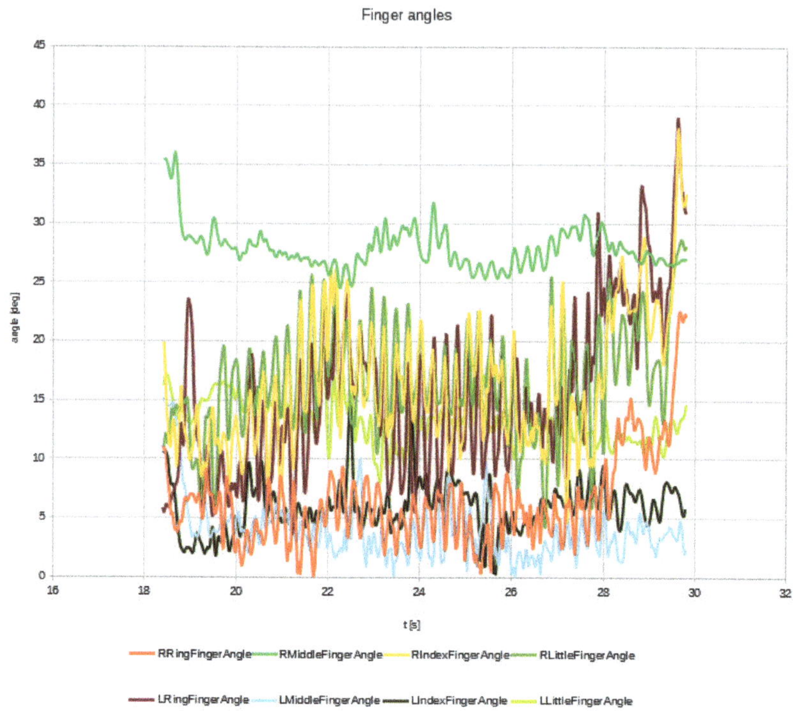

Figure 11. Angles for hands movement for the second sequence.

6. Discussion and Conclusions

The article concerns the first attempt of registering Lazgi, an Uzbek national dance, using an optical motion capture system. For the purpose of this study, a methodology for capturing the movements of this dance was created. Due to the specificity of the Lazgi dance movements performed, the registration of the entire dancer's body using the full Plug-in Gait model was taken into account. This allowed to perform both analysis based on 3D data, represented by each marker attached to the dancer's body, and a kinematic analysis based on the generated outputs of the model used. This set of three-dimensional data gives the possibility of performing a sophisticated analysis of the dancers biomechanics of movements for the successive sequences, rhythm, velocities and others. In order to register the hand movements, which are a crucial element of this dance, a model was developed with which it is possible to analyse the movements of individual dances in detail.

The described methodology was verified by recording the Lazgi dance in a motion capture session. From the obtained data, two dance sequences were selected and analysed. The trajectories of selected markers were determined for both the upper and lower body parts. The left and right movements of the dancer were compared. Additionally, the angles of the upper and lower body parts for the right and left sides were analysed. The first sequence, the bend, is performed with little dynamism. Correctly made, it should reflect the symmetry of movements in two planes. The results presented in Figures 4 and 5 are almost symmetrical with respect to the vertical axis of the body, which proves that the movements were performed correctly. Although the second movements were more dynamic, the axes of symmetry on which the movements were performed can also be seen (Figures 6 and 7). Angle changes for the ankle, knee, pelvis, wrist, elbow and arm were computed for the first sequence. It can clearly be seen that the angles are similar (Figures 8 and 9) and even identical (Figure 8). The presented results of hand movements (Figure 10), including the angles of the fingers (Figure 11), allow to conclude that the model is suitable for the analysis of the Lazgi dance. For a dynamic sequence, the finger movements vary from 0 to 40 degrees (Figure 11).

The obtained results confirm that the proposed methodology is correct for this dance. Moreover, it is universal and can be used both for dances in which the movements performed involve the lower parts of the body (e.g., Greek dances) and the upper parts. In addition, the presented hand model allows to analyse the movements of individual fingers.

Some problems with the presented method should be mentioned. A large number of markers placed on the hands disturbs the course of motion capture by obscuring them. This significantly increases the labour intensity of post-processing.

The obtained data can be used in five aspects: (i) dance analysis (kinematics, kinetics, sequences, rhythm and others), (ii) classification and recognition of movements, (iii) making ICH data public (e.g., in the form of a database), (iv) creating and visualising 3D models and (v) developing software solutions (e.g., VR or AR).

Additionally, the acquired data and the developed dance movement models will be eternally stored and made available on the Internet. This will contribute to the preservation and popularisation of the intangible cultural heritage.

Author Contributions: Conceptualization, M.S.-P., M.M. and E.L.; methodology, M.M., E.M.; software, J.S.; validation, J.S.; formal analysis, J.S., M.S.-P.; investigation, J.S., M.S.-P., E.L., P.P. and D.M.; resources, M.S.-P., P.P., E.M. and M.M.; data curation, J.S., M.S.-P., P.P. and E.L.; writing—original draft preparation, M.S.-P., P.P. and M.M.; writing—review and editing, M.S.-P., P.P. and M.M.; visualization, J.S., M.S.-P.; supervision, M.S.-P., E.L. and M.M. All authors have read and agreed to the published version of the manuscript.

Funding: This article has been partially supported by the Polish National Agency for Academic Exchange under Grant No. PPI/APM/2019/1/00004 titled "3D DIGITAL SILK ROAD".

Institutional Review Board Statement: The study was conducted according to the guidelines of the Declaration of Helsinki, and approved by the Commission for Research Ethics of Lublin University of Technology, No. 2/2016 dated 8 April 2016.

Informed Consent Statement: Informed consent was obtained from all subjects involved in the study.

Data Availability Statement: The dataset created during the study can be found on the website: https://cs.pollub.pl/zasoby/?lang=en (accessed on 23 November 2021).

Conflicts of Interest: The authors declare no conflict of interest.

References

1. Intangible Heritage Domains in the 2003 Convention. Available online: https://ich.unesco.org/en/intangible-heritage-domains-00052 (accessed on 12 October 2021).
2. Pistola, T.; Diplaris, S.; Stentoumis, C.; Stathopoulos, E.A.; Loupas, G.; Mandilaras, T.; Kalantzis, G.; Kalisperakis, I.; Tellios, A.; Zavraka, D.; et al. Creating immersive experiences based on intangible cultural heritage. In Proceedings of the 2021 IEEE International Conference on Intelligent Reality (ICIR), Virtual Event, 12–13 May 2021; pp. 17–24.
3. Lack, K.J. Capturing Dance: The Art of Documentation (An Exploration of Distilling the Body in Motion). Available online: https://ro.ecu.edu.au/theses_hons/77 (accessed on 12 October 2021).
4. Khorezm Lazgi: The Sunniest Dance on Earth. Available online: https://voicesoncentralasia.org/khorezm-lazgi-the-sunniest-dance-on-earth/ (accessed on 12 October 2021).
5. Rakhimov, B. The Eternity of Lazgi. *Eurasian Music Sci. J.* **2020**, *2*, 22. [CrossRef]
6. Khorazm Dance, Lazgi. Available online: https://ich.unesco.org/en/RL/khorazm-dance-lazgi-01364 (accessed on 12 October 2021).
7. Kico, I.; Liarokapis, F. Enhancing the learning process of folk dances using augmented reality and non-invasive brain stimulation. *Entertain. Comput.* **2022**, *40*, 100455. [CrossRef]
8. Laraba, S.; Tilmanne, J. Dance performance evaluation using hidden Markov models. *Comp. Anim. Virtual Worlds* **2016**, *27*, 321–329. [CrossRef]
9. Lee, H.M.; Garg, S.; Lim, K.M.; Lee, H.P. The sound of drumming for traditional Taiko drumming and lion dance. *Appl. Acoust.* **2022**, *185*, 108434. [CrossRef]
10. Chan, J.C.P; Leung, H.; Tang, J.K.T.; Komura, T. A Virtual Reality Dance Training System Using Motion Capture Technology. *IEEE Trans. Learn. Technol.* **2011**, *4*, 187–195. [CrossRef]

11. Han, T.; Yao, H.; Xu, C.; Sun, X.; Zhang, Y.; Corso, J.J. Dancelets Mining for Video Recommendation Based on Dance Styles. *IEEE Trans. Multimed.* **2017**, *19*, 712–724. [CrossRef]
12. Stavrakis, E.; Aristidou, A.; Savva, M.; Himona, S.L.; Chrysanthou, Y. Digitization of Cypriot Folk Dances. In *Progress in Cultural Heritage Preservation*; Ioannides, M., Fritsch, D., Leissner, J., Davies, R., Remondino, F., Caffo, R., Eds.; Springer: Limassol, Cyprus, 2012; Volume 7616 LNCS, pp. 404–413.
13. Kico, I.; Liarokapis, F. Investigating the Learning Process of Folk Dances Using Mobile Augmented Reality. *Appl. Sci.* **2020**, *10*, 599. [CrossRef]
14. Voulodimos, A.; Rallis, I.; Doulamis, N. Physics-based keyframe selection for human motion summarization. *Multimed. Tools Appl.* **2020**, *79*, 3243–3259. [CrossRef]
15. Voulodimos, A.; Doulamis, N.; Doulamis, A.; Rallis, I. Kinematics-based Extraction of Salient 3D Human Motion Data for Summarization of Choreographic Sequences. In Proceedings of the International Conference on Pattern Recognition, Beijing, China, 20–24 August 2018; pp. 3013–3018.
16. Douka, S.; Ziagkas, E.; Zilidou, V.; Loukovitis, A.; Tsiatsos, T. The 3D Motion Capturing Process of Greek Traditional Dance "Syrtos in Three" and a Proposed Method for the Sex Identification of Performing Dancers. *Adv. Intell. Syst. Comput.* **2021**, 1231, 366–374.
17. Ziagkas, E.; Zilidou, V.; Loukovitis, A.; Douka, S.; Tsiatsos, T. Greek Traditional Dances Capturing and a Kinematic Analysis Approach of the Greek Traditional Dance "Syrtos" (Terpsichore Project). *Adv. Intell. Syst. Comput.* **2021**, 1192, 514–523.
18. Rallis, I.; Georgoulas, I.; Doulamis, N.; Voulodimos, A.; Terzopoulos, P. Extraction of key postures from 3D human motion data for choreography summarization. In Proceedings of the 9th International Conference on Virtual Worlds and Games for Serious Applications, VS-Games 2017, Athens, Greece, 6–8 September 2017; pp. 94–101.
19. Rallis, I.; Bakalos, N.; Doulamis, N.; Doulamis, A.; Voulodimos, A. Bidirectional long short-term memory networks and sparse hierarchical modeling for scalable educational learning of dance choreographies. *Vis. Comput.* **2021**, *37*, 47–62. [CrossRef]
20. Rallis, I.; Bakalos, N.; Doulamis, N.; Voulodimos, A.; Doulamis, A.; Protopapadakis, E. Learning Choreographic Primitives Through A Bayesian Optimized Bi-Directional LSTM Model. In Proceedings of the International Conference on Image Processing, ICIP, Taipei, Taiwan, 22–29 September 2019; pp. 1940–1944.
21. Rallis, I.; Protopapadakis, E.; Voulodimos, A.; Doulamis, N.; Doulamis, A.; Bardis, G. Choreographic Pattern Analysis from Heterogeneous Motion Capture Systems Using Dynamic Time Warping. *Technologies* **2019**, *7*, 56. [CrossRef]
22. Doulamis, A.; Voulodimos, A.; Doulamis, N.; Soile, S.; Lampropoulos, A. Transforming intangible Folkloric performing arts into tangible choreographic digital objects: The Terpsichore approach. In Proceedings of the 12th International Joint Conference on Computer Vision, Imaging and Computer Graphics Theory and Applications, Porto, Portugal, 27 February–1 March 2017; Volume 5, pp. 451–460.
23. Protopapadakis, E.; Rallis, I.; Doulamis, A.; Doulamis, N.; Voulodimos, A. Unsupervised 3D Motion Summarization Using Stacked Auto-Encoders. *Appl. Sci.* **2020**, *10*, 8226. [CrossRef]
24. Kitsikidis, A.; Dimitropoulos, K.; Uğurca, D.; Bayçay, C.; Yilmaz, E.; Tsalakanidou, F.; Douka, S.; Grammalidis, N. A game-like application for dance learning using a natural human computer interface. In *International Conference on Universal Access in Human-Computer Interaction*; Springer: Cham, Switzerland, 2015; pp. 472–482.
25. Kitsikidis, A.; Dimitropoulos, K.; Douka, S.; Grammalidis, N. Dance analysis using multiple Kinect sensors. In Proceedings of the 2014 International Conference on Computer Vision Theory and Applications, VISAPP 2014, Lisbon, Portugal, 5–8 January 2014; pp. 789–795.
26. Kitsikidis, A.; Dimitropoulos, K.; Yilmaz, E.; Douka, S.; Grammalidis, N. Multi-sensor Technology and Fuzzy Logic for Dancer's Motion Analysis and Performance Evaluation within a 3D Virtual Environment. In Proceedings of the 2014 International Conference on Universal Access in Human-Computer Interaction, UAHCI 2014, Crete, Greece, 22–27 June 2014; pp. 379–390.
27. Aristidou, P.; Stavrakis, E.; Charalambous, P.; Chrysanthou, Y.; Himona, S.L. Folk dance evaluation using laban movement analysis. *J. Comput. Cult. Herit.* **2015**, *8*, 20. [CrossRef]
28. Molnár, C.; Pálya, Z.; Kiss, R.M. Static Balancing Ability and Lower Body Kinematics Examination of Hungarian Folk Dancers: A Pilot Study Investigating the "Kalocsai Mars" Dance Sequence. *Appl. Sci.* **2021**, *11*, 8789. [CrossRef]
29. Jain, N.; Bansal, V.; Virmani, D.; Gupta, V.; Salas-Morera, L.; Garcia-Hernandez, L. An Enhanced Deep Convolutional Neural Network for Classifying Indian Classical Dance Forms. *Appl. Sci.* **2021**, *11*, 6253. [CrossRef]
30. Nurindiyani, A.; Pramadihanto, D.; Afifah, R. Motion Modeling of Traditional Javanese Dance: Introduction of Javanese Dancer Gesture with 3D Models. In Proceedings of the 2019 International Electronics Symposium (IES), Surabaya, East Java, Indonesia, 27–28 September 2019.
31. Utsugi, A.; Masaya, T.; Takaoka, H. Analysis of Movements of Body Trunk in Japanese Traditional Dance. In Proceedings of the 2017 International Conference on Culture and Computing, Kyoto, Japan, 10–12 September 2017.
32. Lim, Y.K.; Park, J.W. The way to preserve korean intangible cultural assets. In Proceedings of the HCI International 2018—Posters' Extended Abstracts, Las Vegas, NV, USA, 15–20 July 2018; pp. 192–195.
33. Ribeiro, C.; Dos Anjos, R.K.; Fernandes, C. Capturing and documenting creative processes in contemporary dance. *ACM Int. Conf. Proc. Ser.* **2017**, *2017*, 3078041.
34. Hajdin, M.; Kico, I.; Dolezal, M.; Chmelik, J.; Doulamis, A.; Liarokapis, F. Digitization and Visualization of Movements of Slovak Folk Dances. *Adv. Intell. Syst. Comput.* **2019**, *917*, 245–256.

35. Tongpaeng, Y.; Rattanakhum, M.; Sureephong, P.; Wicha, S. Implementing a tool for translating dance notation to display in 3D animation: A case study of traditional Thai dance. In Proceedings of the 30th International Conference on Industrial, Engineering, and Other Applications of Applied Intelligent Systems, IEA/AIE 201, Arras, France, 27–30 June 2017; pp. 22–26.
36. Tongpaeng, Y.; Sureephong, P.; Rattanakhum, M.; Yu, H. Thai dance knowledge archive framework based on Labanotation represented in 3D animation. In Proceedings of the 2nd Joint International Conference on Digital Arts, Media and Technology 2017: Digital Economy for Sustainable Growth, ICDAMT 2017, Chiang Mai, Thailand, 1–4 March 2017; pp. 66–70.
37. Tongpaeng, Y.; Maphon, P.; Sureephong, P. Improving ThaiLabanXML standard for the automatic Thai laban drag drop system. In Proceedings of the ECTI DAMT-NCON 2019—4th International Conference on Digital Arts, Media and Technology and 2nd ECTI Northern Section Conference on Electrical, Electronics, Computer and Telecommunications Engineering, Nan, Thailand, 30 January–2 February 2019; pp. 168–173.
38. Tongpaeng, Y.; Nantapatsaran, K.; Sureephong, P. Interpreting Thai laban notation into ThaiDanceXML. In Proceedings of the 2018 International Conference on Digital Arts, Media and Technology (ICDAMT), Chiangrai, Thailand, 25–28 February 2018; pp. 173–178.
39. Udawatta, L.; Daniels, B.; Hissem, S.; Wivou, J.; Pathirana, P.N.; Nguyen, D.K.; Nguyen, N.D. Motion Capturing of Biomechanical Systems in Traditional Dance using IMU Sensors. In Proceedings of the 2017 International Conference on Electrical and Computing Technologies and Applications, Icecta, Aurak, United Arab Emirates, 21–23 November 2017; pp. 1–5.
40. Aristidou, A.; Shamir, A.; Chrysanthou, Y. Dance Ethnography: Organizing Large Dance Collections. *J. Comput. Cult. Herit.* **2019**, *12* , 1–27. [CrossRef]
41. Sivic, J.; Zisserman, A. Video Google: A Text Retrieval Approach to Object Matching in Videos. In Proceedings of the ICCV 2003—9th IEEE International Conference on Computer Vision, Nice, France, 13–16 October 2003; pp. 1–8.
42. Plug-in Gait Reference Guide. Available online: https://docs.vicon.com/display/Nexus210/Plug-in+Gait+Reference+Guide (accessed on 20 October 2021).
43. Hernandez-Barraza, L.; Yeow, C.-H.; Varela, M. The Biomechanics of Character Types in Javanese Dance. *J. Dance Med. Sci.* **2019**, *23*, 104–111. [CrossRef] [PubMed]
44. Cook, J.R.; Baker, N.A.; Cham, R.; Hale, E.; Redfern, M.S. Measurements of Wrist and Finger Postures: A Comparison of Goniometric and Motion Capture Techniques. *J. Appl. Biomech.* **2007**, *23*, 70–78. [CrossRef] [PubMed]

applied sciences

MDPI

Article

Preliminary Application of the Algorithm Highlighting Petroglyph Patterns

Piotr Wojcicki *, Sylwester Korga and Marek Milosz

Department of Computer Science, Faculty of Electrical Engineering and Computer Science, Lublin University of Technology, Nadbystrzycka 36B, 20-618 Lublin, Poland; s.korga@pollub.pl (S.K.); m.milosz@pollub.pl (M.M.)
* Correspondence: p.wojcicki@pollub.pl

Abstract: One of the greatest problems in the field of cultural heritage is the process of material destruction of monuments. In most cases, the cause is operational or random wear. Such processes can be slowed down or eliminated thanks to the maintenance process. The most dangerous processes of monument wear and tear are those in which historical objects are exposed to environmental impact. Cycles of weather changes have a particularly negative impact on objects of this type. These types of objects are significantly damaged and are the most difficult to scan. The research work included the scanning of such an object and attempts were made to digitise the point cloud. The monument selected for the model laser scanning was a rock petroglyph located on the slope of the Chatkal mountain near the town of Tashkent in Uzbekistan. The article describes the process of digitally recording a historic object with the use of the Artec Eva scanner. The process of data acquisition from points in space and the processes of highlighting petroglyphs are described. The highlighted model was obtained using proprietary software created for this purpose by the authors of the article. As a result, data were saved which would be difficult to save and read using the previously used 3D scanning techniques. The obtained data can be made available to recipients in the form of exhibits in virtual museums.

Keywords: digitisation of monuments; 3D scanning; petroglyphs; algorithms; virtual museum

Citation: Wojcicki, P.; Korga, S.; Milosz, M. Preliminary Application of the Algorithm Highlighting Petroglyph Patterns. *Appl. Sci.* **2022**, *12*, 1660. https://doi.org/10.3390/app12031660

Academic Editors: Yosoon Choi and Yangquan Chen

Received: 11 November 2021
Accepted: 3 February 2022
Published: 5 February 2022

1. Introduction

From time immemorial, all efforts made to protect monuments have been aimed at preserving them for future generations. Traditional techniques, still widely used today in conservation, are essentially based on the use of science and ancient craftsmanship. This traditional approach to cataloguing monuments has many limitations. Each nation loses its national heritage, e.g., historic buildings [1,2]. This is a serious problem because it is related to the preservation of culture, history and memory of the past. In 1992, the Romanesque Fantoft church in Norway was completely destroyed by fire in a few hours. Giotto's frescoes in Assisi were badly damaged in the 1997 earthquake. Hundreds of ancient temples were destroyed in the 2016 Bagan earthquake. In 2019, a fire broke out in the Notre-Dame cathedral which destroyed the temple, damaging the internal monuments and the structural elements of the roof.

Zahi Hawass, the former Egyptian Minister of State for Antiquities, said that the monuments in Egypt would not be able to survive another hundred years [3,4]. He proposed that dedicated specialist offices be set up in the country to register changes, look for solutions and coordinate the reconstruction of the progressive damage. Sudden losses of world cultural heritage sites are particularly tragic and are prompting scientists to develop technologies to reduce the consequences of the loss of valuable monuments. One of such monuments are petroglyphs. Petroglyphs are signs carved in rock, or drawings called carvings. The word petroglyph comes from the Greek language. It is a combination of two words: petros meaning 'rock' and glyphein denoting 'engrave'. Their combined form was first used by the French as pétroglyphe. Petroglyphs are classified as rock art. The engraving can be deep and add depth to the image, but it can also be a barely visible scratch on the top layer of rock. Pictures of this type have been created for millennia, and are engraved to this day. They are used to convey knowledge about the times when they were created: about contemporary culture, achievements in art, science or economy. They are a commemoration of everyday life or a document of unusual events, and have historical, artistic and scientific value. Petroglyphs should be treated as a form of picture writing. They have various forms depending on the time and place of their creation. According to the [5] literature, they were already created by people from the Neolithic era. The oldest petroglyphs are estimated to be 10–12 thousand years old [6]. They can be found on all continents except Antarctica. Large clusters of petroglyphs are found in various places in Africa, Scandinavia, Siberia, the south-eastern part of North America, and in Australia and Asia.

The study of petroglyphs in Uzbekistan began in the 19th century. In the 20th century, over a hundred petroglyph sites were discovered. Most of the petroglyphs in Uzbekistan belong to different periods of history, but they all have different storylines as well as the art of ancient authors who preserved the stone heritage of ancient tribes for future generations. Proper documentation is a key element in minimising the consequences of such catastrophic losses. Traditional methods of documentation, such as drawings and photographs, however, have many disadvantages that lead to inaccuracies and distortions in the documentation. They are usually very time consuming and prone to human error. The advent of digital technologies has opened up new possibilities for the protection of monuments. One of the new digital recording technologies available is 3D scanning. It provides realistic, digital 3D models of scanned monuments. This technology enables the execution of precise three-dimensional digital documentation and creates an opportunity to maintain the heritage in virtual form. The Polish National Heritage Institute promotes the need to conduct more and more laser scans of monuments and create modern documentation [7,8].

Historical monuments often do not have documentation or it is out of date. Classic, analogue forms of documentation of cultural monuments and other spatial historic objects are projections, sketches and views created as a result of mapping. Along with the development of computer techniques, monuments began to be archived in numerical and digital form. Data is saved in the form of vector and raster files. 3D scanning of historic objects is one of the new digitisation technologies that are constantly developed and improved. Spatial scanning of historic buildings allows for the archiving of cultural heritage. 3D laser scanning of monuments has many pragmatic applications. Creating objective, accurate, three-dimensional documentation is invaluable for the conduct of scientific research [9,10]. Due to laser scanning technology, it is possible not only to study the selected phase of the history of the monument, but also easily follow the subsequent stages of changes taking place in the past.

2. Related Works

In their work [11] Matthias Zeppelzauer, Georg Poier, Christian Reinbacher used the method of figure segmentation. The geometry of the monument's surface was used as a point cloud. They used adaptive learning and interactive segmentation which resulted in a smoothed image. The final results of their experiment are based on applying artificial intelligence, which is related to the use of forecasting. Such methods may inaccurately reproduce the monument. Filip Hedman in his work [12] described the digitisation of rites. For this purpose, he used laser scanning. The data obtained from the scanning process was processed using various filters. Using filters, he removed elements surrounding the petroglyph geometry. He also applied various colouring filters to enhance the rock carvings. Additionally, it used a rendering process in which a texture is applied to the monument. In this way, it obtained a very good visual effect and the final effect corresponds to the original. The results of his work show that it is possible to make this type of visualisation, but it is time-consuming. In their work [13], scientists Christian Horn, Derek Pitman, Rich

Potter used the techniques of landscape topographic analysis to archive petroglyphs. They used the technology applied to record and model the relief which has topographic features. This method was developed to assist in the analysis of 3D models. The results obtained in this method depend on the manner of the incident light. They employed the ArcMap software, used to process geospatial data, and applied the geospatial scanning method to scan petroglyphs. The results of their work were filtered by sharpening and smoothing the image. This method turned out to be effective and the results obtained are of high quality.

3. Petroglyphs in Hodjikent, Uzbekistan

The symbolism of petroglyphs has had a deep cultural and religious significance for the societies forming them. In many cases this meaning is known to their contemporary successors. Petroglyphs belong to the group of historic artefacts that are the most difficult to protect against destruction and degradation. This has to do with the rock erosion process, a slow procedure that destroys petroglyphs. The main factors that cause it are the effects of water, wind and changing temperature cycles. Uzbekistan has a temperate climate with a warm and dry variety. In the region of Tashkent (Tashkent), the climate has mountain features. The summers in this area are dry and hot, with average temperatures of 26 °C in the north and 32 °C in the south. During the day, there are periods of strong heat up to 50 °C. Winters, on the other hand, are cool, often frosty and cloudy. The average temperature in January is −10 °C. There are strong drops in temperature at night. There is a climatic double-decker in the mountains. Uzbekistan is a dry country, the average rainfall is 100 to 200 mm per year. The maximum rainfall is in winter and spring, but in winter there is snowfall. The highest rainfall values fall in alpine areas up to 1000 mm per year, however in the mountains it is mainly snowfall. Such conditions adversely affect the behaviour of rock petroglyphs. Over time, petroglyphs become more and more damaged, or even until they disappear completely. Therefore, an attempt was made to preserve such objects in digital form [14,15]. The digitised petroglyphs are referred to by the name of a small village of Hodjikent, located 70 km from Tashkent. The city and its monuments have been included in the UNESCO World Heritage List. Near Tashkent, on the western slope of the Chatkal ridge on the left bank of the Chirchik River, there is a rock with petroglyphs. This place is especially important for the people of Uzbekistan, because here is the "holy spring". It has been a sacred place for centuries. Hodjikent petroglyphs were discovered in 1949–1950 by H. Alpysbaev. Detailed examination of the site revealed that this petroglyph showed the horns of a wild sheep. At that time, only 23 rock carvings from the 1st millennium BC were discovered. Further research on this monument was continued by the famous Central Asian archaeologist, specialist in large petroglyphs, M. Khuzhanazarov. Research has revealed new petroglyphs that have been carefully studied and added to the existing ones. The petroglyphs were carved on the rock at a height of 12.5 m and a length of 18 m. For many years, the rock engravings were exposed to the influence of the surrounding environment. Due to climatic conditions, some of the rock carvings have not survived. Less visible figures were discovered along with distinct patterns. It is likely that the petroglyphs were made at different times. More than 90 petroglyphs have been discovered in Hodjikent. Some of them depict single characters: ibex, argali, reindeer, horse, ox and dog [16,17]. The most dangerous for petroglyphs are micro cavities caused by material freezing in winter. One of the main damaging factors is water, in the case of rock materials, the more porous the rock material, the more dangerous its impact. When it freezes, water increases in volume and exerts pressure on the walls of the pores and micro-spaces. These influences are usually complex and cause damage to the individual surface layers and their gradual degradation. The specificity of Hodjikent petroglyphs is that the surface of many of them is smooth. Some scientists have suggested that this was due to the touch of pilgrims to the sacred stones over the centuries, but there is a more rational explanation for this phenomenon. According to the scientist archaeologist M. Khuzhanazarov, the whole thing is in the technique of making petroglyphs. Most likely, the authors carved the paintings in stone, and then carefully polished them. We can presume that the technique of their production extended their service life. A sharp rise in temperature is as dangerous as freezing, especially for brittle materials with different coefficients of thermal expansion. The heated surface expands, causing cracks and flaking parallel to the surface.

4. Materials and Methods

An Artec Eva 3D scanner (Figure 1) was used to scan the petroglyphs in Hodjikent. The Artec Eva is a versatile, high-speed 3D scanner. It is characterised by fast and accurate measurement, precise texture and colour reproduction, high resolution 3D scan and ergonomic design. The Eva scanner has been designed with its user in mind—it is light, does not require the use of positioning markers (glued on the scanned object) or calibration, which significantly improves the comfort of work. The scanner performs scanning using structured light, which makes it completely safe for the scanned monuments. In addition, the recording of 16 frames per second allows to perform quick and accurate scans. The basic parameters of the scanners used are presented in Table 1.

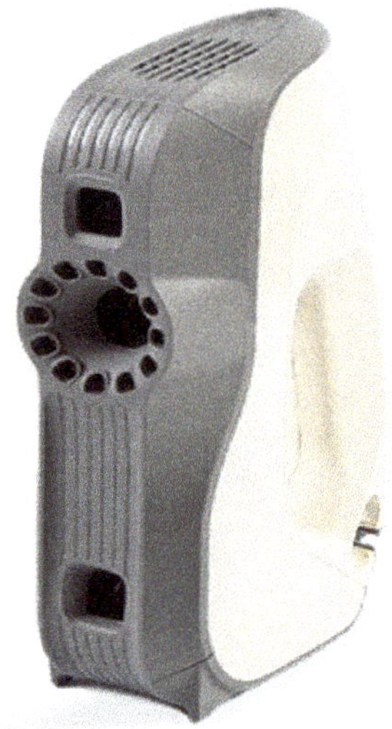

Figure 1. Scanner view: Artec Eva.

The data obtained after scanning were processed using the Artec Studio version 12 Professional. This is software dedicated to Artec scanners.

Table 1. Basic parameters of the Artec Eva scanner.

General Specification	Artec EVA
Working distance	0.4–1 m
Measurement area volume	61,000 cm^3
Linear field of view H × W (min)	214 mm × 148 mm
Linear field of view H × W (max)	536 mm × 371 mm
Angular field of view H × W	30° × 21°
3D resolution, up to	0.5 mm
Accuracy of the 3D point, to	0.1 mm
3D accuracy at a distance, up to	0.03 % at 1 m
Ability to Capture Texture	Yes
Texture Resolution	1.3 mp
Colours	24 bpp
Reconstruction 3D, up to	16 fps
Data download speed, up to	2 million points per second
Structured light source	flash bulb (without laser)

The petroglyph was subjected to a scanning process, Figure 2. An Artec Eva scanner was used.

Figure 2. Petroglyph scanning process.

The Artec Studio version 12 Professional was used for data processing. The first phase of data processing was loading the scanned objects and placing them in three-dimensional space. The loaded objects are shown in Figure 3.

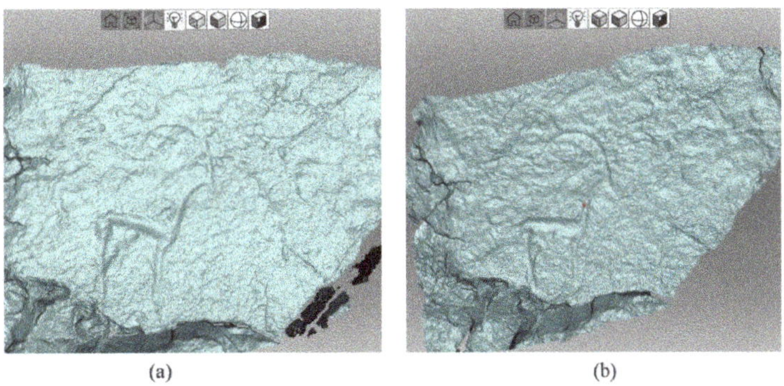

Figure 3. Sample objects in the form of a petroglyph scan, (**a**) scan no. 1, (**b**) scan no. 2.

The superimposition of the scanned objects and their processing made it possible to obtain a 3D mesh model of the petroglyph. A number of activities based on specialised tools were used to build the mesh:

- Registration—Fine registration.
- Edition of scans—Removed artefacts; (selection tools were used, such as: Lasso selection, Cutoff-plane selection and 3D selection).
- Alignment of scans—scans were matched by specifying 3–4 pairs of characteristic points on the scans.
- Registration—Global registration.
- Fusion—Outlier removal.
- Fusion—Smooth fusion.
- Postprocessing—if there were holes, they were repaired using the Hole filling or Fix holes tools.

- Postprocessing—Mesh simplification.
- Applying texture.

The mesh model of the petroglyph generated by the program is shown in Figure 4.

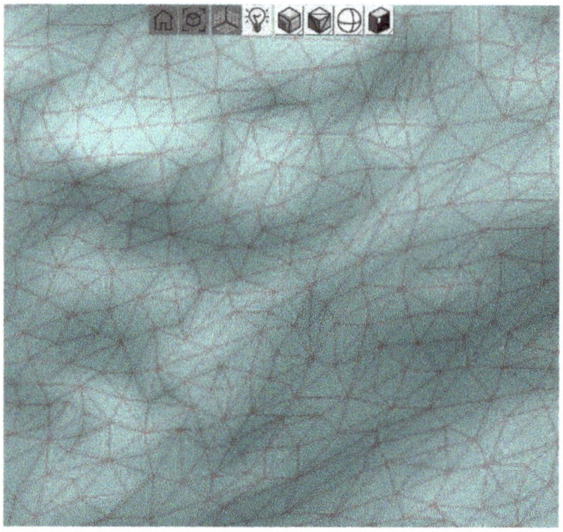

Figure 4. An example of a mesh area of a scanned petroglyph.

The mesh model obtained from the scans had the so-called zero thickness. This was due to the fact that the object (petroglyph) was scanned only from one side—from the front. Moreover, the obtained 3D mesh model would be burdened with defects of mesh geometry. Autodesk's 3ds Max program was used for further processing of the mesh model. With the right tools and Boolean logic operations, the missing areas of the mesh were supplemented and a fully valuable virtual object of the scanned petroglyph was obtained, Figure 5.

Figure 5. View of the scanned object with a rock structure overlaid.

The created object was saved as files with extensions STL and XYZ. The examined petroglyphs were made by subtractive methods. Apart from the drawings themselves, there are many cavities in the investigated rock material. Scanning the surface provides a good quality topographic model, which has the additional consequence of transferring to digital form the grooves. The scanned objects are characterised by high surface roughness, which additionally makes it difficult to find the desired drawing. Due to the rough surface, it is not easy to identify only man-made fragments. It is assumed that the drawing is continuous, i.e., it has connection between successive fragments of the pattern being searched for. The processing consisted of preparatory stages which included:

1. Reading of the mesh model.
2. Assessment of the location of the pattern in relation to the floor plane (*XY*).
3. 3D transforms of point clouds (rotations around the axis *X*, *Y*, *Z*).
4. Removing unnecessary fragments of the model.
5. Sorting points on the *y* coordinate.
6. Creating a section for a given interval $m < y < n$.
7. Representation of the cross-section in the diagram.
8. Searching for local minima with specific properties for each cross-section (width and depth of the 'well').
9. Joining sections—the vicinity of the points found must be within the range (elimination of local grooves).

Algorithm 1 shows how to deal with the coordinate file (*XYZ*).

Algorithm 1: Cloud data pre-processing and cross-section creation

Input: Coordinate file (x, y, z).
Parameters: Parameters defining the cross-section creation step, window parameters.
Output: Coordinate file (x, y, z).
1: Estimation of the model redundancy and the position of the pattern surface in relation to the floor plane (XY).
2: Removing unnecessary fragments of the 3D model.
3: Transforms of point cloud (rotations around the axis OX, OY, OZ).
4: Sorting points on the y coordinate.
5: Creating a cross-section for a given range (m, n).
6: Searching for local minimum with specific properties for each cross-section.
7: Joining sections—the vicinity of the points found must be within the range.

The first step of the presented algorithm is to visualise the file containing the coordinates of each scanned point using the Ovito—Open Visualization Tool [18]. Preliminary evaluation of the visualized point cloud allows to determine whether a transformation around any of the *X*, *Y* or *Z* axes is required in order to arrange the petroglyph surface parallel to the *XY* plane. Another reason may be the need to reduce the point set for faster processing [19]. The next step is to remove the model fragments. Reducing the number of points shortens the processing time and speeds up acquisitions of the final solution. The selection of fragments is manual and allows to reduce the size of the model by about 70%. If a deviation is found, transformations are carried out in accordance with Formulas (1)–(3).

$$R_X(\alpha) = \begin{bmatrix} 1 & 0 & 0 & 0 \\ 0 & cos\alpha & -sin\alpha & 0 \\ 0 & sin\alpha & cos\alpha & 0 \\ 0 & 0 & 0 & 1 \end{bmatrix}, \quad (1)$$

$$R_Y(\alpha) = \begin{bmatrix} cos\alpha & 0 & sin\alpha & 0 \\ 0 & 1 & 0 & 0 \\ -sin\alpha & 0 & cos\alpha & 0 \\ 0 & 0 & 0 & 1 \end{bmatrix}, \quad (2)$$

$$R_Z(\alpha) = \begin{bmatrix} cos\alpha & -sin\alpha & 0 & 0 \\ sin\alpha & cos\alpha & 0 & 0 \\ 0 & 0 & 1 & 0 \\ 0 & 0 & 0 & 1 \end{bmatrix}, \quad (3)$$

where R is the elementary rotation matrix in the *X*, *Y*, and *Z* axes, respectively, and α is the rotation angle.

$$\begin{bmatrix} x' \\ y' \\ z' \\ 1 \end{bmatrix} = R \cdot \begin{bmatrix} x \\ y \\ z \\ 1 \end{bmatrix} \tag{4}$$

The coordinates of the points after the rotation around any axis will be the result of multiplying the specific R matrix by the coordinates of each of the original points. Alignment in this way allows to generate a cross-section that can be easily interpreted, because the coordinates of each successive layer do not differ significantly from the previous one. After the initial preprocessing operations, the coordinates are saved to the local database. The use of a database enables the execution of several steps of the implemented algorithm simultaneously. The database search operation is more efficient than the operation with large text files because of the ability to perform both filtering and sorting with one command. When the data are loaded into the database, the stage of selecting the model fragments with a given step according to the relationship begins:

$$\bigwedge_{x_i, y_i, z_i \in P} \bigwedge_{m < y_i < n} P_s = (x_i, z_i), \tag{5}$$

where x_i, y_i, z_i are coordinates of the i point, and m and n are cut-off values for a specific plane. Using an interval instead of a single value was considered, because only a few points could then meet the condition. Finally, a larger set of points belonging to several planes satisfying the inequality $m < y < n$ was analysed. A point was selected with coordinates that meet, firstly, within the examined set of points, and secondly, the y coordinate for a given set is in the range (m, n).

For each data set meeting the above relationship, the coordinates of a given point are grouped to a dependence that performs the $z(x)$ function. The section created in this way is subjected to further analysis.

In the next step, the process of searching for at least one local minimum with specified criteria is started. The problem is not trivial due to the existing roughness and grooves visible as minor changes in the value of the z coordinate for subsequent x.

The local minimum on each of the planes is detected using a window of l values. Successive values of $z(x)$ are checked and if a significant decrease is detected, and then a sufficiently large increase, the anomaly is considered as a minimum.

After completing the search for the minimum for a single cross-section, the coordinates of points for the layer belonging to the interval are saved (5). According to Algorithm 1, the minimum search is repeated for each layer.

After all the slices are searched for the local minimum, the proximity of these extremes on the k nearest layers is checked. The comparison of k sets of coordinates follows and in the case of a difference in the values of x within the window, a neighbourhood is found. If there is no neighbourhood, the minimum coordinates are removed from the list. This solution ensures that small grooves that are definitely not part of the pattern you are looked for are omitted. The ready set of coordinates extracted from the above algorithm is removed from the original cloud set.

5. Results

The data obtained in accordance with the description in Section 4 were used for the study.

5.1. Visualisation

Visualisation allows to determine whether and where the object of interest is located in the model. In addition, it allows to assess whether it is necessary to carry out a reduction, e.g., when the model is very large and the object sought is relatively small. An example of a complete set of points is presented in Figure 6.

Figure 6. Sample set of petroglyph points.

5.2. Model Reduction

The reading time of the point cloud as well as its processing depends on the size, therefore the first step after visualization is to reduce the size. It is obvious that it is impossible to get rid of the set of points located in the area of interest. Therefore, this step is completely manual. The reduced model is shown in Figure 7.

Figure 7. The model after removing unnecessary elements.

5.3. Matrix Transformations

In order to generate a cross-section on the basis of which it is possible to find the local minimum (grooves), the model pattern should be placed parallel to the XY plane. Then each subsequent cross-section will not differ diametrically from the previous one. Using Formulas (1)–(4), all positions of the points should be recalculated in such a way as to achieve the parallelism of the planes. The above example clearly shows a deviation in the X axis. The angle of rotation is selected empirically. The model before and after the rotation around the X axis is shown in Figure 8.

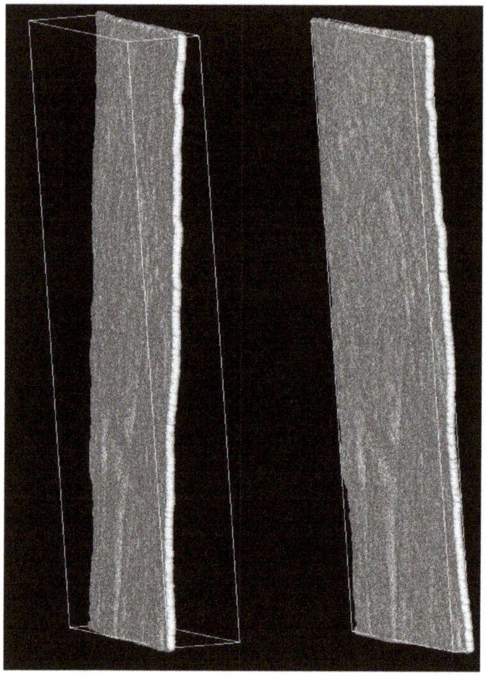

Figure 8. The model before and after rotation around the X axis.

5.4. Cross-Section

After the model is properly positioned, the cross-section is created along the Y axis. Each of the cross-sections is a relationship that consists of several layers of the Y direction. This is due to the fact that for a given y coordinate there are not enough points to determine the local minimum. An example of a cross-section is shown in Figure 9. The points selected from the set must therefore belong to the interval that satisfies the inequality (5). Figure 9 clearly shows two drops in values which are responsible for the engraved 'legs' of the goat shown on the tested rock.

5.5. Searching Minimum

The search for a local minimum is a key element of the work presented. It is assumed that in each of the cross-section layers there must be at least one minimum in the pattern region. The lack of a local extreme is interpreted as the lack of a pattern in a given layer, similarly an occurrence on a single section only. The algorithm analyses k consecutive z values in the window according to the x-order. When a decrease and an increase are detected, the data are temporarily saved to the table. After the search is completed, for a given section, local minima of the specified width and depth are selected, while others are ignored. The extreme dimensions must be consistently repeated on successive layers in the vicinity of detection. Such a procedure allows to reduce the risk of recognising a natural groove as part of the pattern. In the case of extensive natural engravings located on the surface the algorithm may consider them as part of the pattern.

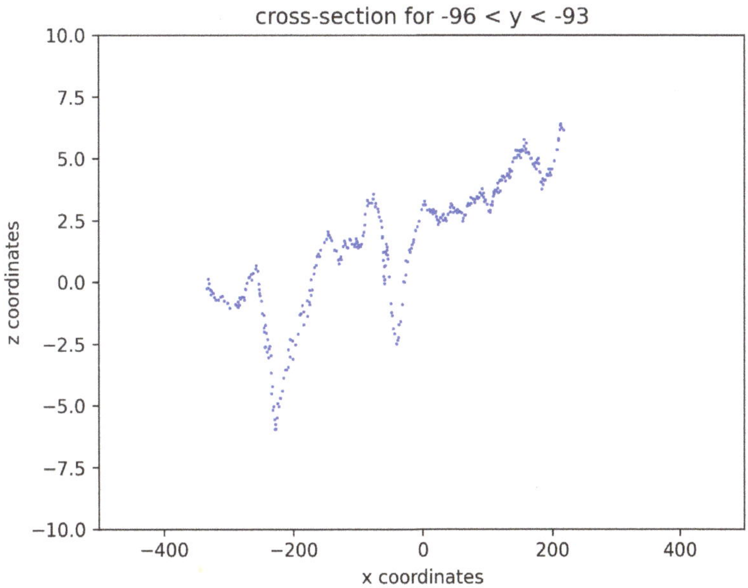

Figure 9. The model cross-section for $y \in (-96, -93)$.

5.6. Pattern Removal

For the obtained set of points classified as an element of the formula, the operation of subtracting from the main model is performed. The highlighting of the petroglyph is done by iterative removal of points. The removal takes place in several iterations, which causes points to be mainly removed from the vicinity of detected extremes. As a result, an in-depth model was obtained, which shows human activity more clearly than before. The effect of the algorithm's operation is shown in Figure 10.

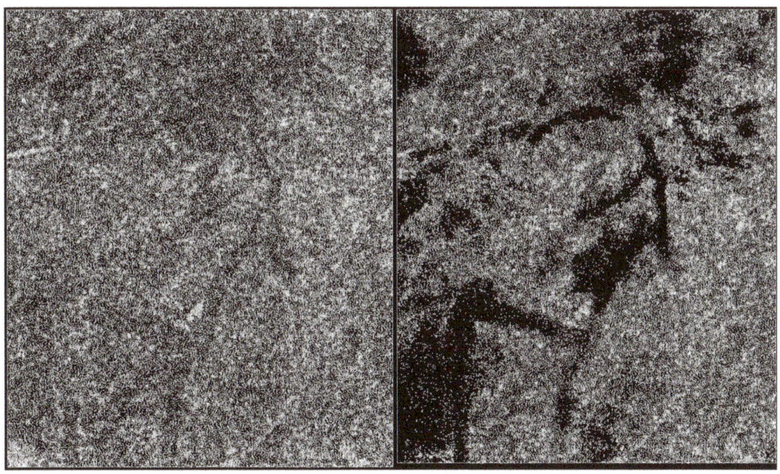

Figure 10. The model before and after highlighting the detected pattern.

Extracted points belonging to the pattern can be successfully used to create a new model that is a negative for a rock pattern (cast), e.g., for use in 3D printing. After applying the algorithm, the place and shape of the pattern being searched for was clearly shown, which was difficult to see before.

6. Discussion

The 3D scanning process of petroglyphs allowed for detailed mapping of the object in the virtual world. The 3D spatial inventory provides a complete set of data on the monument, which in turn can be used for further work. It is an excellent solution for scientists, archaeologists and historians who would like to conduct research. The prepared 3D model of the object will be perfect as a basis for creating reconstruction or revitalisation projects. There are many reasons why 3D scanning should become a widely used method of documenting in the protection of monuments:

- 3D scanning poses no risk to scanned objects. 3D scanners look like cameras and the equipment never touches the object or emits harmful radiation. This technology deserves the name of a fully non-invasive method of examining historic buildings.
- 3D scanning offers an extremely accurate representation of the object. Mapping unusual and complex shapes with the help of documentation made by hand would be impossible. In addition, 3D scanners are able to document the colours and textures of the scanned surfaces, which makes them useful also in the case of the presence of paint layers, which is very important in the case of petroglyphs.
- While scanning petroglyphs, data related to the surrounding space, e.g., the shape of rocks, is simultaneously collected [20].
- 3D scanning is fast and efficient, especially compared to traditional measurement and documentation methods. Making a 3D model of a petroglyph, without laser scanning, but only on the basis of photos and traditional measurements, also involves a huge expenditure of time and work. However, the time needed to obtain a point cloud of even a very complex object is several minutes [21,22].
- 3D scanning of petroglyphs is considered to be the most accurate monument recording technology that we can create today.
- Virtual objects can be used as exhibits to create virtual museums. These types of exhibitions are increasingly used in the field of museology.
- 3D scans of monuments and cultural heritage are an invaluable aid for museums that want to present their exhibits online. Regardless of their command of the relevant language or their knowledge, recipients can discover the common cultural heritage in an interesting and modern way from anywhere in the world.
- Digital mapping of historic objects also gives the opportunity to assess the level of their damage. This may prove invaluable especially in the case of monuments that have been destroyed or devastated.
- The 3D scans were suitable for further processing and were not rejected by the software due to too many errors.
- Processing data from a point cloud to obtain a mesh model of a petroglyph using Artec Studio 12 is a time-consuming operation and requires quite a lot of computing power [23].
- The model highlighting algorithm can be used for preparatory operations, e.g., before 3D printing. It allows to better visualise the pattern in the digital model.

In the models used, the number of points in the cloud originally exceeded 250 and 300 thousand (file sizes: 10.8 MB and 12.6 MB), respectively; after processing it was reduced to about 70 thousand (reduction rate is over 70%). An interesting functionality of 3D scanning processes is also modern education and promotion, where visualisations of historic architecture in virtual reality are used to disseminate knowledge about monuments. The development of knowledge and active participation in getting to know the cultural heritage contributes to the protection of monuments for future generations [24].

7. Conclusions

The use of 3D scanning technology allows to easily transfer any real object to the virtual three-dimensional world. Due to the high detail of measurements, it can also be used to carry out 3D building inventories. Recent years have seen the development of the field of museology, where virtual museums are increasingly used. Interestingly, they are becoming more and more popular. No technology can replace personal contact with works of art. Personal contact with a monument is often related to the atmosphere of the surroundings and the people of a given region. Scanned historic objects can be recreated as replicas using 3D printing technology. Such replicas of monuments are created for blind people who cannot touch the monument and get to know it. 3D scanners are perfect for archaeological works [14,25]. It is possible to carry out with them extremely accurate scanning of earth layers at archaeological sites. Removing layers of earth and working with monuments irretrievably destroys their arrangement. The way the objects are arranged in the ground gives a lot of information about the find. A digital record of such posts helps to save this information for return [26]. The paper presents the preliminary application of the algorithm highlighting petroglyph pattern. The presented example shows the point cloud processing and the final result. Future works assume the improvement of the algorithm in terms of better detection of points belonging only to the pattern being searched and omitting the remaining objects.

Author Contributions: Conceptualization, M.M. and S.K.; methodology, M.M., S.K. and P.W.; software, P.W. and S.K.; validation, M.M., S.K. and P.W.; formal analysis, S.K.; investigation, P.W.; resources, M.M.; data curation, P.W.; writing—original draft preparation, P.W., S.K. and M.M.; writing—review and editing, M.M., S.K. and P.W.; visualization, S.K.; supervision, M.M.; project administration, M.M. All authors have read and agreed to the published version of the manuscript.

Funding: This article has been supported by the Polish National Agency for Academic Exchange under Grant No. PPI/APM/2019/1/00004 titled "3D DIGITAL SILK ROAD".

Institutional Review Board Statement: Not applicable.

Informed Consent Statement: Not applicable.

Data Availability Statement: The dataset created during the study can be found on the website: https://cs.pollub.pl/zasoby/?lang=en (accessed on 16 December 2021).

Conflicts of Interest: The authors declare no conflict of interest.

References

1. Beckensall, S.; Laurie, T. *Prehistoric Rock Art of County Durham, Swaledale and Wensleydale*, 1st ed.; CBD: Beckenham, UK, 1998.
2. Beckensall, S.; Laurie, T. *Prehistoric Rock Art in Northumberland*, 1st ed.; CBD: Beckenham, UK, 2001.
3. Hawass, Z. International Journal of Cultural Property 4.1. In *The Egyptian Monuments: Problems and Solutions*; UNESCO: Paris, France, 2007; pp. 154–196.
4. Hawass, Z. Can the Sphinx Be Saved? *Archaeology* **1994**, *47*, 42–43.
5. Amenta, N. *Surface Reconstruction Encyclopedia of Algorithms*; Springer: New York, NY, USA, 2008; pp. 1–5.
6. Ronald, M. *The Prehistoric Rock Art of Galloway and the Isle of Man*, 1st ed.; Blandford Press: London, UK, 1979.
7. Kęsik, J.; Montusiewicz, J.; Żyła, K. Low-cost 3D Scanning Methods as Tools for Virtual Heritage Expositions. *Meas. Autom. Control* **2014**, *60*, 664–667.
8. Pereira, F.G.; Martínez, J.C.; Lourenço, A.B. Virtual Restoration of a XVIII Century Sculpture. *Res. Cent. Sci. Technol. Arts* **2009**, *3*, 4–13.
9. Castango, R. Il Restauro si fa in 3D Fablab di Parma: "Le Imprese ci Aiutino". La Rivoluzione Dello Spazio 3D. 2015. Available online: http://parma.repubblica.it (accessed on 29 October 2021).
10. Kęsik, J.; Montusiewicz, J.; Kayumov, R. An approach to computer-aided reconstruction of museum exhibits. *Adv. Sci. Technol. Res. J.* **2017**, *11*, 87–94. [CrossRef]
11. Zeppelzauer, M.; Georg, P.; Seidl, M.; Reinbacher, C.; Schulter, S.; Breiteneder, C.; Bischof, H. Interactive 3D Segmentation of Rock-Art by Enhanced Depth Maps and Gradient Preserving Regularization. *J. Comput. Cult. Herit.* **2016**, *9*, 19. [CrossRef]
12. Hedman, F. Improved Visualization of Rock Carvings. Bachelor's Thesis, Uppsala University, Uppsala, Sweden, 30 November 2016.

13. Horn, C.; Pitman, D.; Potter, R. An evaluation of the visualisation and interpretive potential of applying GIS data processing techniques to 3D rock art data. *J. Archaeol. Sci. Rep.* **2019**, *27*, 13. [CrossRef]
14. Comes, R.; Buna, Z.; Badiu, I. Creation and Preservation of Digital Cultural Heritage Journal of Ancient History and Archeology. *World Sci. Publ.* **2008**, *1*, 150–156.
15. Berndt, R.; Buchgraber, G.; Havemann, S.; Settgast, V.; Fellner, D. Publishing Workflow for Cultural Heritage Artifacts from 3D Reconstruction to Internet Presentation. In *Digital Heritage, Proceedings of the Euro-Mediterranean Conference, Lemessos, Cyprus, 8–13 November*; Springer: Berlin/Heidelberg, Germany, 2010; pp. 166–178.
16. Li, R.; Luo, T.; Zha, H. 3D Digitization and Its Applications in Cultural Heritage. In *Digital Heritage, Proceedings of the Euro-Mediterranean Conference, Lemessos, Cyprus, 8–13 November*; Springer: Berlin/Heidelberg, Germany, 2010; pp. 381–388.
17. Pavlidis, G.; Koutsoudis, A.; Arnaoutoglou, F.; Tsioukas, V. Methods for 3D Digitization of Cultural Heritage. *J. Cult. Herit.* **2007**, *8*, 93–98. [CrossRef]
18. Stukowski, A. Visualization and analysis of atomistic simulation data with OVITO—The Open Visualization Tool. *Model. Simul. Mater. Sci. Eng.* **2010**, *18*, 015012 [CrossRef]
19. Sitnik, R.; Mączkowski, G.; Krzesłowski, J. *Calculation Methods for Digital Model Creation Based on Integrated Shape, Color and Angular Reflectivity Measurement*; Springer: Berlin/Heidelberg, Germany, 2010; pp. 13–27.
20. Toshew, Y.E.; Hieu, L.C.; Stefanova, L.P.; Tosheva, E.Y.; Zlatov, N.B.; Dimov, S.T. Reverse Engineering and Rapid Prototyping for new orthotic devices. In *Intelligent Producion Machines and Systems*; Elsevier: Amsterdam, The Netherlands, 2005.
21. Montusiewicz, J.; Milosz, M.; Kayumov, R. 3D digital technologies in the practical training of archaeologists. In Proceedings of the 10th International Technology, Education and Development Conference (INTED 2016), Valencia, Spain, 7–9 March 2016; pp. 7451–7460.
22. Liy, Y.; Pottmann, H.; Wang, W. Constrained 3D shape reconstruction using a combination of surface fitting and registration *Comput. Aided Des.* **2006**, *38*, 572–583. [CrossRef]
23. Raja, V.; Fernandes, K.J. Reverse Engineering. In *An Industrial Perspective*; Springer: London, UK, 2008.
24. Montusiewicz, J.; Czyż, Z.; Kayumov, R. Selected methods of making three-dimensional virtual models of museum ceramic objects. *Appl. Comput. Sci.* **2015**, *11*, 51–65.
25. Patay-Horvath, A. The virtual 3D reconstruction of the east pediment of the temple of Zeus at Olympia an old puzzle of classical archaeology in the light of recent technologies. *Digit. Appl. Archaeol. Cult. Herit.* **2014**, *1*, 12–22. [CrossRef]
26. Ramandeep, S.; Ashish, S.; Sneh, A.; Britty, B. Validation of reverse-engineered and additive-manufactured microsurgical instrument prototype. *Surg. Innov.* **2016**, *23*, 606–612.

applied sciences

MDPI

Article

Good Practices in the Use of Augmented Reality for the Dissemination of Architectural Heritage of Rural Areas

M. J. Merchán [1], P. Merchán [2,*] and E. Pérez [2]

[1] Facultad de Educación, Universidad de Extremadura, 06006 Badajoz, Spain; mjmerchan@unex.es
[2] Escuela de Ingenierías Industriales, Universidad de Extremadura, 06006 Badajoz, Spain; emilianoph@unex.es
* Correspondence: pmerchan@unex.es

Abstract: When it concerns dissemination, the relationship between cultural heritage and technology has not always been easy. On the one hand, the emotions involved in knowing, enjoying, and feeling a real heritage remain should not be eclipsed by the technological experience. On the other hand, technology cannot be relegated to the point where its use is so superficial as to make it irrelevant. Hence, in the search of good practices in the dissemination of cultural heritage, it is essential that interdisciplinary work teams, composed of humanists and technicians, design new experiences that try to achieve a balance between what must be shown and the way in which it is done. This paper aims to show the potential of augmented reality (AR) for the dissemination of cultural heritage, since it allows the perceiving of real remains while offering virtual reconstructions and complementary information, sounds, images, etc. A bibliometric analysis focusing on its use for the preservation of the memory of vernacular architecture is made. Then, from the understanding and interpretation of the ideas exposed in literature, a decalogue of good practices in the use of AR for architectural heritage is established considering the holistic perspectives that any dissemination initiative must have. These initiatives can attract people to rural areas and contribute to the improvement of their social and economic situation. As an example, a proof of concept of an AR experience is presented at the end of the paper.

Keywords: IT in cultural heritage; augmented reality; 3D modelling; cultural heritage dissemination; modern exhibition techniques

Citation: Merchán, M.J.; Merchán, P.; Pérez, E. Good Practices in the Use of Augmented Reality for the Dissemination of Architectural Heritage of Rural Areas. *Appl. Sci.* **2021**, *11*, 2055. https://doi.org/10.3390/app11052055

Academic Editor: Marek Milosz

Received: 27 January 2021
Accepted: 20 February 2021
Published: 25 February 2021

1. Introduction

One of the main problems that administrations of developed countries are facing today is the depopulation of rural areas. Concerning this, the concept of "empty Spain" has emerged in our country to define those regions with a worrying rural population drift. The "emptied" territories (both Castile and Extremadura, mainly) have alarmingly low population densities in a society that is growing older. In these territories, countless villages and small towns have been abandoned since the middle of the last century, and many others have been worryingly losing population for decades.

The emigration of peasants that began in the mid-20th century, the generalisation of higher education that gave rise to new specialists that the villages were unable to absorb, and the low birth rates suffered in modern Western societies were the main factors that provoked the economic collapse of these areas, the abandonment of places, and the maladministration of the countryside. In recent years, rural tourism, that is, city people seeking the tranquillity and simplicity of rural areas, has partly alleviated the problem of the abandonment of villages. However, in general this tourism is a very seasonal and short-term, and does not contribute at all to creating new economic structures that foster the permanence of inhabitants in the place, hiding the real problems of rural areas: isolation (poor roads or low internet connectivity), lack of supplies and services, the ageing of the population, etc.

Despite this modern return to the "locus amoenus", the word "rural" still has a pejorative meaning, and the belief that customs and ways of life, speech, or even architecture are inferior, backward, and crude, is hard to eradicate. This fact, maintained over the centuries, has led to the destruction of part of the rural architectural heritage for the sake of misunderstood modernity.

Vernacular Architecture

The term "vernacular" is the most widespread when defining this kind of rural architecture. Its etymology alludes to what is representative of a specific place and time. Thus, it is an architecture characterized by its dependence on local needs, resources, and traditions, which makes it impossible to copy without losing its essence. It is an architecture without architects, since it has incorporated the skills and experience of local builders [1]: it is a "resilient" architecture that knew how to adapt to the different circumstances and needs that arose [2].

Vernacular architecture has been revisited in recent years by architectural specialists in search of the energy and resource efficiency it has always boasted, however, the consideration of these attributes has not always been so. As mentioned previously, in the case of Spain, the abandonment of the countryside in the mid-20th century caused many of the buildings to collapse due to a lack of use and conservation. The arrival of democracy entailed a period of economic prosperity, which brought, in turn, a change in production and consumption models, along with an increase in population. These changes meant that both the old spaces associated with agricultural and livestock farming and the dwellings near them underwent remodelling to adapt them to new production and housing needs (Figure 1), which meant their complete disappearance in most cases. Only in places where the economic boom was not enough to reactivate the population and to change the ways of life in a profound manner did the old buildings continued to survive.

Figure 1. "La bomba", in 1985 and nowadays. Old troughs and houses, Fuente del Maestre, Spain (Photo: Manuel García).

Paradoxically, the fact that some places have become stuck in the past has turned them into a tourist attraction. Consequently, in some of these villages, it has been possible to restore their most unique buildings, giving them a second lease of life and saving them from disappearance. However, this is not a widespread trend. In some areas, far from the usual tourist routes, this heritage is still being lost nowadays. This situation has led some administrations (local, regional, and national) to start adopting actions to alleviate the problem of disappearance. One of the most common resources is the creation of repositories of vernacular architecture, as is the case of the Junta de Extremadura [3]. However, this is not a guarantee of survival. Unfortunately, a comparison of the images taken at the time of cataloguing and currently reveal that a gradual and severe deterioration is occurring.

Therefore, the preservation of heritage, both cultural and natural, tangible and intangible, is an inescapable task for today's society. Heritage is the identity of peoples and preserves collective memories, helping to reinforce the idiosyncrasy of places and to lay the foundations for the future. Cultural heritage, in particular, must be a source of inspiration and social integration, and also an aid to boost the economic dynamism that increases the capacity for rural revitalisation [4].

On the path towards the total protection of heritage buildings, the use of technology has allowed specialists in conservation and preservation to develop new multidisciplinary methods that help them to improve results and optimise time [5]. Among the most widely used techniques, those related to scanning and 3D modelling, as well as the development of virtual and augmented reality applications, are the most promising [6]. These methods can be implemented to ensure the physical survival of the heritage good [7] or to keep its memory when, unfortunately, it is condemned to disappear [8]. Similarly, technology has become an essential tool when disseminating and making immovable heritage known. It can be considered as the best vehicle to connect young users with their territory and its cultural value through the daily use of mobile devices.

As known, unlike VR where the user interacts in a fully recreated world, AR is concerned with generating layers of virtual information that must be correctly aligned with the image of the real world to achieve a desired sensation. Thus, AR is situated between real and virtual environments, and is responsible for constructing and aligning objects that are integrated into a real scenario. It allows users to see the real world in real time, enhanced with virtual object information once this is coordinated correctly in order to display a coherent superposition. The application of AR aims to enhance real-site experiences through the interaction of digital contents blended with the real environment through the screen of a digital device [9]. Recent advancements in the capabilities and affordability of hand-held devices suggest that AR content has reasonable accessibility.

Accordingly, in recent years, educational facilities, such as museums and galleries, have increasingly made use of the benefits offered by AR to enhance the learning experience of visitors. Similar trends have also been observed in architectural heritage, where the idea of applying AR to its study, conservation, and dissemination has emerged strongly.

Following this line of thought, this paper analyses the use of augmented reality to help people know, understand, and value vernacular architecture. The AR experiences designed for such purposes must not only ensure the resilience of the buildings but also contribute to the socio-economic revitalisation of the places in which they are located. To this end, Section 2 begins with a bibliometric analysis addressing the scientific activity in the use of AR for the dissemination of cultural heritage, continuing with a review of the state-of-the-art practices on this topic, which includes articles from the bibliometric analysis but covers a wider spectrum of the literature in this respect to give a broader view and better contextualisation. As a result of the analysis of the literature, A decalogue of good practices on the use of AR for the dissemination of vernacular architecture as a way of reactivating the economy of rural areas is stated in Section 3. The second part of this section is dedicated to briefly presenting the procedure we followed to develop an AR application to visit an archaeological site that had to be covered to allow a road to pass over it. Section 4 consists of a succinct discussion of the paper. Finally, some conclusions are outlined in Section 5.

2. Materials and Methods

In order to establish a decalogue of good practices, we considered it necessary to first review the experiences published to date. The analysis of this review will allow us to reach a series of conclusions from which to draw the decalogue.

2.1. Bibliometric Analysis

As said, with the aim of analysing and studying scientific activity in the use of AR for the dissemination of cultural heritage, we have carried out a bibliometric analysis on the Scopus database. To retrieve data, we performed the following queries (on 17 December 2020): topic = (TITLE-ABS-KEY ("augmented reality" AND "cultural heritage")); topic = (TITLE-ABS-KEY ("augmented reality" AND "cultural heritage" AND "dissemination")); topic = (TITLE-ABS-KEY ("augmented reality" AND "architectural heritage")); topic = (TITLE-ABS-KEY ("augmented reality" AND "architectural heritage" AND "dissemination")).

The total number of documents found for the first query (relating to "augmented reality" with "cultural heritage") was 693. This number was reduced to 46 when the term "dissemination" was added. There were only 25 documents for the query with "architectural heritage" and this number dropped to 5 when "dissemination" was added to the search. Figure 2 shows two plots with the number of documents sorted by year. The first aspect to highlight is that this is a relatively recent field of research (the first document appears in 1999), in which activity has clearly increased in recent years, reaching a peak in 2019 with a total of 118 papers for "cultural heritage". When the term "dissemination" is added to the search, the maximum value also happens in the same year. Besides, it can be seen that the first works that relate to both concepts are not published until 2005. Later still does the association between "augmented reality" and "architectural heritage" occur, and it is not until very recently that the "augmented reality" was used for the dissemination of architectural heritage, as seen in Figure 2b.

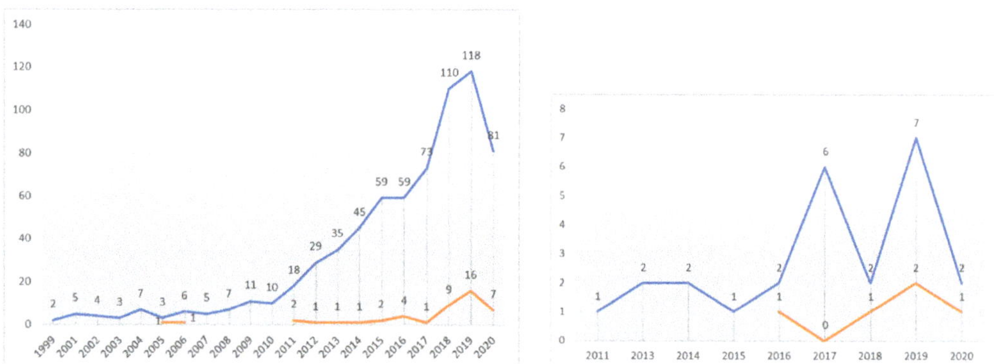

(a) (b)

Figure 2. Life cycle of the publications relating "augmented reality" with (**a**) "cultural heritage" (in blue) and "dissemination" (in orange); (**b**) "architectural heritage" (in blue) and "dissemination" (in orange).

Table 1 shows the type of document for each query in percentages. In all cases the largest number of documents corresponds to publications at conferences, except for documents dealing with augmented reality applied to the dissemination of architectural heritage.

Table 1. Type of documents (%).

Doc. Type	Cultural Heritage	Cultural Heritage Dissemination	Architectural Heritage	Architectural Heritage Dissemination
Conference Paper	63.6	50.0	52	20
Article	24.3	39.1	36	80
Conference Review	5.4	2.2	4	-
Book Chapter	4.3	4.3	4	-
Review	1.6	2.2	4	-
Book	0.3	2.2	-	-
Editorial	0.3	-	-	-
Letter	0.1	-	-	-

When analysing the number of documents per subject area, it is worth pointing out that the contributions from the fields of computer sciences, engineering, and mathematics are the most numerous for "cultural heritage". In the query concerning "cultural heritage" and "dissemination", the first field is still computer sciences, but social science and arts and humanities become the second and third most important fields. These three fields are also the first when searches incorporate the terms "architectural heritage", and "architectural heritage " and "dissemination" (Figure 3).

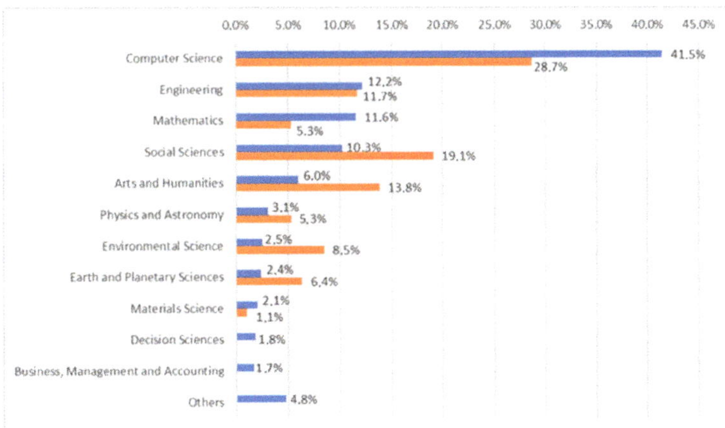

(**a**)

Figure 3. *Cont.*

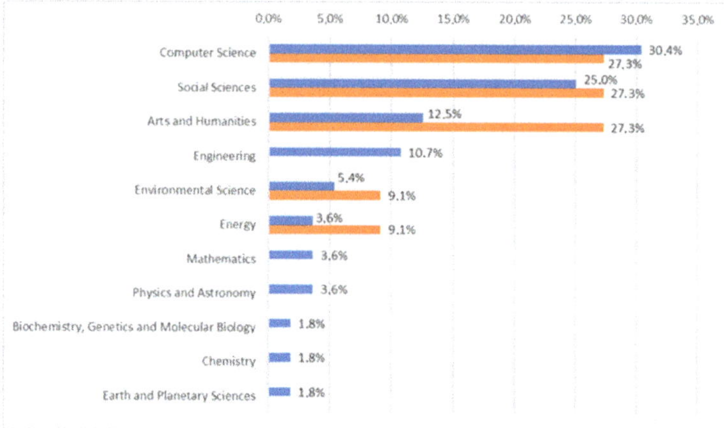

(b)

Figure 3. Subjects of the publications relating "augmented reality" with (**a**) "cultural heritage" (in blue) and "dissemination" (in orange); (**b**) "architectural heritage" (in blue) and "dissemination" (in orange).

Finally, as shown in Figure 4, it is interesting to note that most of the production is European, with Italy clearly at the top of all the queries made in the database. It is also important to highlight the third position of Spain, along with Greece, in the queries concerning "cultural heritage" and the second position of Spain, together with Canada, Colombia, and Malaysia, for the queries concerning "architectural heritage", which gives an idea of the level of interest in these fields that exists in the authors' country.

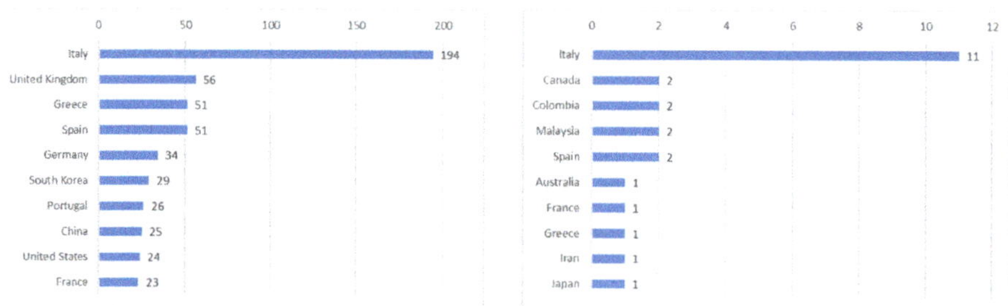

(a) (b)

Figure 4. Top ten countries of the authors of the publications relating "augmented reality" with (**a**) "cultural heritage" and (**b**) "architectural heritage".

2.2. Augmented Reality for Cultural Heritage Dissemination: A Review

After the bibliometric analysis of the publications in the field was completed, we reviewed their content with the idea of identifying features that AR applications must have to fulfil the objective of disseminating cultural heritage.

This review of the wider spectrum of publications (the one that relates "augmented reality" with "cultural heritage") reveals two trends in terms of the applicability of AR to this field. On the one hand, the older theories, which are more limited since the possibilities of the technology were not fully developed, consider these types of experiences in terms of their use either as guides to visit museums and historical sites or for a better appreciation/knowledge of a work of art as, for instance, in [10–15]. The use of AR to guide visits to cultural heritage sites is still in use today, as can be seen in S11, S18, S19 (listed in Table 2), since AR has demonstrated itself to be a good resource for improving the visitor experience in museums and historical sites.

Table 2. Documents selected from the query "augmented reality" and "architectural heritage".

Paper	Reference
S1 [16]	Calisi et al., Architectural historical heritage: A tridimensional multilayered cataloguing method. IISPRS Archives, 2011, 38 (5W16), pp. 599–606.
S2 [17]	Meschini et al., Disclosing Documentary Archives: AR interfaces to recall missing urban scenery. Proceedings of the DigitalHeritage 2013, 2, art. no. 6744783, pp. 371–374.
S3 [18]	Durand et al., Ray-on, an on-site photometric augmented reality device. Journal on Computing and Cultural Heritage, 2014, 7 (2), art. no. 7.
S4 [19]	Rossi et al., A framework to increase the video-mapping accuracy of an architectural heritage mock-up. ACM International Conference Proceeding Series, 2014 April, art. no. a3.
S5 [20]	Cordido et al., The altarpieces of de la Candelaria in Guarenas and del Rosario Chapels in San Jacinto de Caracas Church: A graphic method to establish their historic relation, 2016, Revista 180, 38.
S6 [21]	Cárdenas, M.I.Z., Augmented reality application for the dissemination of cultural heritage [Aplicación en realidad aumentada para divulgación del patrimonio cultural]. Kepes, 2016, 13 (14), pp. 33–59.
S7 [22]	Azmin et al., Architectural heritage restoration of Rumah Datuk Setia via mobile augmented reality restoration. Planning Malaysia, 2017, 15 (1), pp. 139–150.
S8 [23]	Petrucci et al., Musepick: An integrated technological framework to present the complex of Santissima Annunziata in Ascoli Piceno (Italy). ISPRS Archives, 2017, 42 (5W1), pp. 557–564.
S9 [24]	Albourae et al., Architectural heritage visualization using interactive technologies. ISPRS Archives, 2017, 42 (2W5), pp. 7–13.
S10 [25]	Ioannidi et al., Flaneur: Augmented exploration of the architectural urbanscape. Proceedings—IEEE Symposium on Computers and Communications, 2017, art. no. 8024582, pp. 529–533.
S11 [26]	Morganti et al., Habanapp: Havana's architectural Heritage a click away. ISPRS Archives, 2018, 42 (2), pp. 723–730.
S12 [27]	Banfi et al., Extended reality and informative models for architectural heritage: From scan-to-bim process to virtual and augmented reality. Virtual Archaeology Review, 2019, 10 (21), pp. 14–30.
S13 [28]	Germanese et al., Architectural heritage: 3D documentation and structural monitoring using UAV. CEUR Workshop Proceedings, 2019, 2320, pp. 1–12.
S14 [29]	Shabalina et al., Investigating regional heritage through the development and playing of AR games. Proceedings of the European Conference on Games-based Learning, 2019 October, pp. 631–638.
S15 [30]	Luigini et al., 3D digital models for a widespread museum: The Renon's "Bauernhöfe", ISPRS Annals of the Photogrammetry, Remote Sensing, and Spatial Information Sciences, 2019, 42 (2/W9), pp. 447–453.
S16 [31]	Matini et al., A 3D approach to reconstitution of the adobe citadel of Bam after earthquake. International Journal of Architectural Heritage, 2019, 13 (4), pp. 600–618.
S17 [32]	Rashid et al., Geospatial platforms and immersive tools for social cohesion: The 4D narrative of architecture of Australia's Afghan cameleers. Virtual Archaeology Review, 2020, 11 (22), pp. 74–84.
S18 [33]	Li et al., Development of an augmented reality application for protecting the perspectives and views of architectural heritages. IEEE International Conference on Consumer Electronics—Taiwan, ICCE Taiwan 2020, art. no. 9258299.
S19 [34]	Lin et al., In-depth learning of architectural heritage with application of augmented reality based on sequential scenes. IEEE International Conference on Consumer Electronics—Taiwan, ICCE Taiwan 2020, art. no. 9258120.
S20 [35]	Templin et al., The use of low-cost unmanned aerial vehicles in the process of building models for cultural tourism, 3D web and augmented/mixed reality applications. Sensors (Switzerland), 2020, 20 (19), art. no. 5457, pp. 1–26.

On the other hand, the latest studies contain up to five diversifications of the possible uses of AR for cultural heritage, including uses that are shared with other immersive technologies (virtual reality and mixed reality). These five categories are education, exhibition improvement, exploration, reconstruction, and virtual museums, and were systematised in [6]. Obviously, the previous trend is included within this one, but now from a broader and more enriching perspective.

In accordance with the concept of this work, focusing on architecture within the broad concept of cultural heritage, we have elaborated Table 2 with the resulting documents for the query "augmented reality" and "architectural heritage". Not all 25 documents are listed because some of them were compilations or reviews and so the analysis to extract the information summarized in Table 3 was not applicable to them. Thus, Table 3 identifies the categories mentioned in the previous paragraph in which each of these documents can be included.

Table 3. Identification of the categories in which the papers in Table 2 can be included.

Reference	Guide	Work of Art	Exhibition	Education	Reconstruction	Virtual Museum
S1	×	×	×	√	√	√
S2	×	×	×	×	√	×
S3	×	×	×	×	√	×
S4	×	×	√	√	×	×
S5	×	√	×	×	√	×
S6	×	×	×	√	√	×
S7	×	×	×	×	√	×
S8	×	×	×	×	√	√
S9	×	×	×	×	√	√
S10	√	×	×	√	×	×
S11	√	×	×	×	×	×
S12	√	×	√	×	√	×
S13	×	×	×	×	√	×
S14	×	×	×	√	√	×
S15	×	×	×	√	√	√
S16	×	×	×	×	√	√
S17	√	×	√	√	√	√
S18	√	×	×	×	×	×
S19	√	×	×	×	×	×
S20	√	×	√	×	×	√

In the following lines, the main aspects to consider when designing an AR experience, that is, "how" (technological aspects) and "for what" (application), will be examined through a reflective revision of the bibliography on the theme.

To begin with, there are several technology-related issues that need to be solved. Needless to say, the better the nature of their resolution, the better the experiences offered will be. This is especially true for applications dealing with complex and unstructured environments such as archaeological sites. Therefore, the proposal of new methods that solve some of the technological problems existing today will improve both AR systems in general and the outcome of any given situation in particular. This is hence an open field for researchers. Among the technological aspects that can be addressed [6], the most interesting are:

- The modelling of virtual environments;
- The tracking the location of users;
- The 3D registration of real images with virtual models.

As far as 3D modelling is concerned, different techniques can be used depending on the objective of the application being developed. In the simplest cases, the use of 3D design software which allows the creation of artificial virtual models, such as 3DS Max or Blender, suffices. When the application to be performed involves the reconstruction of heritage assets, it is necessary to use other tools to generate 3D models from the real data. The most common techniques used to acquire 3D data are photogrammetry (SfM) [36] and 3D laser scanners [37]. In this case, a new problem must be faced: the adaptation of these 3D models for real-time visualisation. Some proposals regarding resolving this problem have been published recently, although it is not yet totally solved. Undoubtedly, this is an important issue whose resolution will improve the quality of AR applications in the field of cultural heritage [38].

Tracking the location of users allows one to know where they are. The 3D registration determines the users' points of view in order display on the devices used for this purpose the virtual models superimposed on the actual heritage pieces. It is also responsible for modifying the orientation and scale of the virtual models so that they are correctly seen on the real image. Both problems, tracking and registration, are related to each other and are solved by combining information from different sensors in most cases: GPS, inertial measurement units (IMU), and cameras. The information provided by the two first types of sensors is usually utilised to track the location. Meanwhile, the information obtained from the application of computer vision algorithms on the images captured by the cameras is commonly used to locate the users, although it can also be used to solve registration problems. In the case of AR applications in outdoor heritage sites, GPS or IMUs, along with cameras, are the most used sensors. For indoor heritage sites, a camera alone is usually used to locate the user and to solve the registration problem through mark recognition [39,40]. Even so, the problem of real-time registration is still an open issue only solved for the simplest geometric shapes. Obtaining the same real-time results for free-forms and organic objects as those of cultural heritage is far from achieved. In AR applications, it is essential to avoid latency, since the delay makes the experience unsatisfactory for users, despite the perfect technical results.

When designing any system or software that involves a non-expert user, it is very important to follow user-centred design (UCD) processes [41,42]. This is particularly important in the cases discussed in this paper, as the applications designed are intended to be used by a wide range of people. In other words, in general, it must be assumed that most users have no previous technical (mobile and AR) or historical knowledge. A wide variety of tools and methodologies for research, design, rapid prototyping, and user testing of mobile AR user experiences can be found in the literature (e.g., [43]).

Apart from the interface, within the design process both the amount and the flow of information the user should receive has to be decided. It is obvious that too much information could overwhelm the user. In this sense, three parameters can be considered when designing AR and VR applications: presence, affordance, and usability.

Presence is usually defined as "the degree to which a user feels like they are in reality while experiencing the extended reality". In Barfield's study [44], some parameters for measuring presence are provided. An example can be found in [45], where it is reported that some users "lose sight of where the game ends and reality begins". Therefore, in an application, the higher the level of presence, the better the result of the impact and assimilation of the information in the user. Of course, it is very difficult to quantify this, and no generalizations can be made.

Another concept to consider is affordance, which can be defined as "a factor inducing an action for a user to accurately recognize and operate the meaning of an object of interaction in an augmented reality environment". To measure this parameter, some items can be extracted from Harton's classification of the types of affordance [46].

Finally, the concept of usability is also essential, redefined as "the degrees of effectiveness, efficiency, and satisfaction for a user completing a task in a augmented reality environment" in [47]. The studies of Anderson and Shapiro [48], Nielsen [49], and Dünser [50], provide very valuable experience in the evaluation of usability.

As for the works in Table 2, most of them neither explicitly mention nor explain any aspect related to the presence, affordance, or usability of the application they develop. Some of them do highlight the ease of use, given their orientation to the general public attending museums or visiting historical buildings. Only in S6 are both presence and usability are analysed, and in S10 just usability. S20 highlights the increase in presence due to the use of augmented reality devices. On the contrary, in S7 they refer to the poor visualization and the delay offered by the augmented reality experience presented.

2.3. Augmented Reality for Cultural Heritage Enhancement

As previously mentioned, one of the applications of AR in cultural heritage should be facilitating the research of archaeologists, historians, and conservation experts. However, history is a quite immovable field of knowledge since these specialists are usually reluctant to abandon old methods. This could be the reason why it is very difficult to introduce the innovations that technology provides to enhance the investigations in these areas. Even so, some new initiatives have appeared in the last decade, which mean a big leap forward with regards to the previous way of working. Unfortunately, in the case of AR, literature focused on this topic is quite rare [51]. Experts have always considered it as a means of entertainment and enjoyment and a way of enabling people to approach cultural heritage. Thus, needs surrounding the use of AR as a tool for research have not been consolidated yet.

The other two aspects to be taken into consideration when talking about cultural heritage are education and dissemination, which are connected. In this case, education plus dissemination means preservation. There is no better way to preserve a heritage building than to know it, appreciate it, and put it to good use.

One of the most common applications of AR systems, and one of the oldest, is the facilitation and improvement of user experiences IN interpretation/visitor centres and museums. This is usually done by adding signs in real images, which, for example, may indicate the itinerary of the exhibition or show extended information about pieces on display [52,53]. Another digital advance, based on the high-end rendering capabilities of game engines and the AR applicability, is the development of animated pedagogical agents which play the role of guides or educators to guide visitors and explain the exhibitions [54]. Other relational agents can be found in Traum et al. [55]; Lane et al. [56]; or Vosinakis and Avradinis [57].

It is also worth mentioning the use of AR to show reconstructions of cultural heritage assets that cannot be visualised because they have completely disappeared (Figure 5) or to display the missing parts of buildings in ruins. In the latter case, AR allows the visual integration of the reconstructed/modelled parts with the real ruins that still exist [58–60].

The last point to mention is the use of AR to create "serious games". This type of game is designed for educational purposes with the aim of supporting players to accomplish learning objectives in a fun way [61]. To date, these games are mainly focused on formal education. However, informal educational settings can also benefit from "serious games". When used for the dissemination of cultural heritage, they are a perfect way to attract new types of visitor, such as children and teenagers [62].

(a)

(b)

Figure 5. Photos of the archaeological site "La Matilla": (**a**) aerial view of the exterior area; (**b**) interior of one of the three ovens located at the site.

3. Results

As already mentioned, cultural heritage assets in general, and vernacular architecture in particular, help to generate identity, values, and culture, propitiate cohesion, and even to support community experiences. The importance of this sort of architecture goes beyond this, as it can also be the focus of a tourist attraction, becoming an economic engine in many previously depressed areas. Moreover, experts from the public and private sectors consider it a driver for local development, as it is a powerful source of income. This change of approach to the value of vernacular architecture, from considering its conservation and maintenance as a burden to understanding it as an important asset in society, must be carried out together with educational actions and successful practices. However, sometimes it is not easy to persuade certain types of people to invest in technological proposals, as is the case with AR experiences, over others that are considered a priority. All initiatives undertaken must be supported by empirical evidence and practical demonstration of their impact on the economic, cultural, and social improvement of the surrounding area. This can only be achieved through intelligent management [4].

Therefore, any proposal related to obtaining economic and social benefits from the vernacular architecture assets should not be made without taking into account the inhabitants of the area. They are the "owners", and they should be the only beneficiaries. In this case, they need to be convinced that the medium-term benefit is great. To this end, it is equally important that they are engaged and committed in the creation of the AR experience.

The economic viability of this type of project is reinforced by the possibility of AR to be run on visitors' mobile devices (smartphones and tablets). This makes the investment in inventory material and maintenance quite low. Most of the funding may be used for data acquisition, design, and implementation of the AR application. In the documentation phase, it is essential to count on the collaboration of the inhabitants, so that they feel involved in the project and can contribute their knowledge about the buildings and their ideas to improve the AR application. One of the disadvantages we could face when implementing an experience like this is the lack of internet connectivity that some rural areas suffer from. Proposing an eye-catching experience for the authorities could encourage investment in this sense. The capacity to involve stakeholders who sponsor this type of project would also be another factor in achieving its viability.

Bearing in mind the people at whom this type of initiative is targeted, we must also consider a number of determining factors: the experiences must have the capacity to attract a wide and varied audience, from the elderly to families with young children and from the inhabitants of the surrounding areas to specialists in any of the subjects involved in creating them. Likewise, they must create in users the desire to visit the location and stay to continue seeing and knowing it.

The initiative can be considered a success if, on the one hand, it serves as a tourist attraction and, on the other hand, it provides stable economic benefits to the community concerned, either indirectly (trade, hotel nights, etc.) or with the creation of jobs associated with its implementation.

3.1. Decalogue of Good Practices

Although some books, papers, and documents on best practices for the intelligent management and dissemination of cultural heritage assets in general and historical buildings in particular have been published recently [63], they are manuals mainly inspired by the progress of cities and urban architecture [64–66]. To the best of our knowledge there is nothing specifically focused on the rural environment so far.

On the basis of these manuals and the premises mentioned above, as well as on the basis of the initiatives gathered in the reviewed literature, we are in a position to outline some good practices that should inspire the creation and implementation of AR experiences that are beneficial for the socio-economic environment in which they are developed. The main factors that motivated us to compose this decalogue were the necessities for acquiring, cataloguing, and visualising historical buildings exposed in S1; the empirical experience performed in S6 that provides a summary of methodological actions in its Table no. 2; the methodology used in S7 to systematise the five steps to follow (presented in the flow diagram in its Figure 2); the solutions found in S14 from the perspective of gamification, which fosters the attraction and implication of the public with the AR application; and the idea of social cohesion worked on in S17 to engage autochthonous people and visitors with the remains and the experience.

The task that should be undertaken first and foremost is to convince both local authorities and the general population of the need to create such strategies to reach and attract different types of tourists. Once the need has been created, we consider it essential to proceed as follows:

1. Document the heritage assets under study, their state of conservation, location, accessibility, etc.
2. Make a socio-economic study of the town and its surroundings, communication routes and connectivity.
3. Evaluate tourist flows, typology, tastes, seasonality, etc.
4. Study the economic viability of the proposal, assessing both the costs of staff, material, equipment, maintenance, etc., and the direct and indirect benefits.
5. Look for investors, both public and private, to finance the project.

Once the previous points have been acknowledged, in terms of the creation and implementation of the AR experience, these five practices are considered the best:

6. Technology must be used as a means, not as an aim. It is used to improve the knowing, learning, and enjoyment of the building.
7. Internet connectivity must always be available and the necessary facilities for the normal development of the experience assured.
8. The AR application has to be free for visitors, easy-to-use, and able to run on any average device. When designing the experience, ensure the desired levels of presence, affordance, and usability. Improvements, updates, further developments, etc., must be guaranteed for two years after the start of the experience.
9. The information provided by the AR experience must be easily understandable. At the same time, it should create the expectation for the user that they will continue to learn more about the surroundings.
10. The implementation of the AR experience must be complemented with a good communication strategy through social networks, institutional announcements, presentations at fairs, conferences, etc., to reach the largest possible audience.

3.2. An Example: AR Application of the Archaeological Site "La Matilla"

We are developing an AR application for the dissemination of cultural heritage in a particular case in which the site is no longer accessible. Specifically, the application is intended for the visualization of some archaeological remains of Roman buildings and ovens that were found by chance during refurbishment works of the road EX-360, near the town of Fuente del Maestre (Badajoz, Spain), at the place known as "La Matilla". Some photos of the site can be seen in Figure 5. After analysing the findings, evaluating their historical value, and calculating the increase in time and cost that their conservation would entail for the road works, the authorities decided to cover them once digitized. This would at least provide the raw material to generate tools that allow people to visualize these remains later on, when already hidden, either by physical reproduction or by representation through digital technologies: screens, VR, or AR devices. Up to now, we have developed a proof of concept of an AR application to enable the visualization of the site with a mobile phone.

When we decided to design this AR application for the remains found, we took into account that although Fuente del Maestre (Badajoz) is a little town of less than 7000 inhabitants it has a very important cultural heritage. Its archaeological, historical, and monumental wealth led the Government of Extremadura to declare it an "asset of cultural interest with a historical set category" (Decree 136/1998 of 17 November). Due to this, we had at our disposal very interesting bibliography, news, and documents on the surroundings to fulfil points 2 and 3 of the decalogue. Obviously, the data acquisition of the remains and its historical and archaeological study (point 1 of the decalogue) was made before this AR application were conceived.

Concerning the point 5, the local authorities showed interest in the project and will study the budget of the initiative once this AR application became a reality. Besides, this town is home to two of the most important export companies in Spain, which employ a significant number of people and are a focus of attraction for many sales agents, workers, businessmen, politicians, etc. The implication of these private firms will be the next step.

All these advantages motivated us to undertake the design of the AR application thanks to the funding provided by the research projects in which we participate, which for now complies with the point 4 of the decalogue in this phase of "development" we are in.

In our application, technology becomes the only means to make accessible the remains that had to be buried. On the one hand, AR allows researchers to analyse this type of historical construction and, on the other hand, it gives the general public the opportunity to know and explore, above ground level, remains that are buried at a certain depth beneath their feet. This application will undoubtedly be a tool that will allow both target audiences to enjoy the building and learn about these characteristic remains (point 6 of the decalogue).

The idea is to install a fixed stand, suitably signposted, in the area where the buried remains are located. This signage would consist of a poster that would include an explanation of the process of downloading and installing the app, a user manual, and information with summary photographs of the site for users who prefer not to install the application.

Fortunately, currently, mobile internet networks have improved considerably in terms of both coverage and speed. Specifically, it has been verified that in the area where the buried archaeological site is located, internet connectivity is available (point 7 of the decalogue).

Our application is intended to be always free of charge. As mentioned above, these remains are located in a region rich in heritage and which has strong support from the authorities to promote initiatives that study, protect, and disseminate historical heritage. Therefore, it is to be expected that there will always be backup for maintenance of the application so that it is always available to visitors (point 8 of the decalogue).

To develop the proof of concept we have so far, we followed the procedure shown in Figure 6.

The 3D digitization process consisted of the combination of three different technologies: laser scanning, structured light scanning, and digital photography. Since each of them produces data with different resolutions, we decided the following strategy to concurrently apply them in the way that was fastest and most optimal:

- The laser scanner was employed to digitize the exterior area of the site, producing 3D high resolution coloured point clouds.

- The structured light (handheld) scanner was used to acquire the interior of the three ovens, producing 3D high resolution textured meshes.
- The digital camera was utilized to take hundreds of photos of both the exterior area of the site and the interior of the ovens, for later use in photogrammetry software. After the computation, 3D low resolution textured meshes were generated.

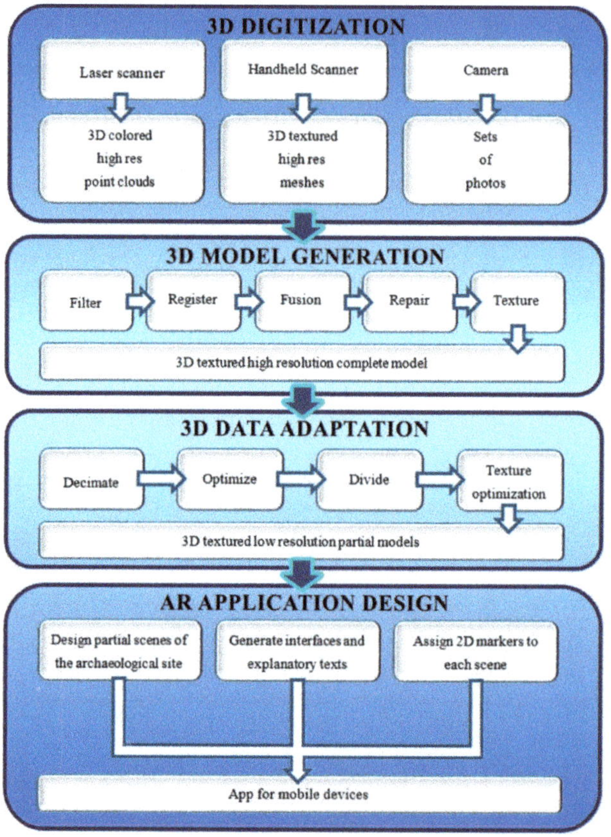

Figure 6. Scheme of the procedure to produce the proof of concept of the AR application.

Then, the next stage in the procedure is the processing of all the data obtained separately from each capturing technology to generate the 3D model of the site. We can summarize this step by enumerating and briefly describing the sequence of tasks that compose this process:

1. Filter: eliminate undesired data and errors.
2. Register: apply the appropriate transformations (position, rotation, scale) to make all the sets of 3D data accurately fit to one another.
3. Fusion: generate a unique mesh with all the information provided by the digitization technologies.
4. Repair: analyse and correct the errors of the mesh to generate a manifold mesh.
5. Texture: compose a unique texture map with optimized distribution from all the texture information generated by the different digitization technologies used.

At this point, we have a digital, accurate 3D model, useful in a great variety of applications. Regarding the application of AR, since it is intended to run on mobile devices, both the resolution of the 3D mesh and the texture must be reduced. This leads to a third stage in our procedure: 3D data adaptation. It consists of the following tasks that provide the results shortly described below:

1. Decimate: drastically reduce the number of points of the mesh.
2. Optimize: slight reduction of the number of points optimizing their distribution, concentrating the points where curvature is greater.
3. Divide: split the mesh into smaller parts to be visualized in different scenes of the AR application. The reason for doing this is due to the technological limitations of mobile devices when representing big 3D objects. This task is related to both the decimation and optimization performed in the previous tasks.
4. Texture optimization: To alleviate the processing of 3D data visualization in mobile devices it is also important to compose the texture map in the most optimal way.

After this stage, the 3D data are prepared to be integrated in the AR application. Finally, the application itself is implemented. A simple way to explain its development process is by dividing it into the following tasks:

1. Design partial scenes of the archaeological site. In our case, we had five 3D-optimized models as a result of the previous stage: two corresponding to the external area and three corresponding to each of the ovens. These models are now enhanced by adding ornamental 3D models or by including background sounds, resulting into five different scenes.
2. Generate interfaces and explanatory texts. It is important to design a usable application that provides the adequate material to help people improve their knowledge, learn, and enjoy, and that generates the need for further information about the site. Interfaces must be intuitive since it will be used by a very diverse group of people (points 8 and 9 of the decalogue).
3. Assign 2D markers. One of the methods currently available to launch the 3D representations in the AR applications is the use of 2D markers. These markers are basically 2D images which are recognized by the application to begin the visualization of a 3D model. In our proof of concept, we assigned one image with a text for each of the 3D models. Figure 7 depicts two of the markers.
4. Generate the application for the mobile devices. Since we have used Unity software to create the AR application, it is relatively easy to reorient the developed content to be compiled for different types of hardware. So far, we have only generated an application for Android mobiles.

Figure 7. Two of the markers designed to be used in the AR application.

Figure 8 shows some images of the use of our AR application and a view of the interior of one of the ovens that can be visited in the experience.

(**a**) (**b**) (**c**)

Figure 8. Two moments of the use of our AR app: (**a**) the user focuses the camera on the marker; (**b**) the 3D model of the site is shown over the image of the marker; (**c**) the 3D model of one of the ovens displayed in the AR application.

Regarding the level of presence (point 8 of the decalogue), it is undeniable that AR applications for cell phones do not offer a presence comparable to the immersion offered by a VR equipment. In any case, although the 3D scenarios have had to be optimized for mobile devices, the quality of the textures produces a very realistic effect in the visualization of the site, which definitely contributes to increasing the level of presence in the experience.

Since the application is still in the prototype phase, it has not yet been disseminated in conferences, social networks, or through institutional announcements (point 10 of the decalogue). This is an issue that is considered of great relevance in order to achieve the application's own objective: raise awareness of the remains, particularly in the case where people may not know of their existence, and, at the same time, make these people participants in the revolution that the application of technologies for the dissemination of heritage, the learning about history, and the digital preservation of valuable remains that have survived the passage of time entails. In the medium term, we plan to first launch a local information campaign among the inhabitants of the site's surroundings and then to extend this campaign to the whole region.

4. Discussion

Throughout these pages, we have discussed about the problem of "empty Spain", the causes of the depopulation, and the consequences this fact has at present. Despite this, these empty areas have a rich architectural heritage of traditional buildings. This vernacular architecture, which we have also defined in this paper, can be one of the main economic resources for the places where they are located, provided that the necessary strategies are developed for its conservation, enhancement, and dissemination. However, it is not always easy for local authorities to undertake this task.

Technology has come to alleviate some of these problems by facilitating the work of heritage experts. By the same token, technology can also be an economic driver for these depressed regions to regain their vitality.

Among other alternatives currently available, AR can be used as a tool for the communication and dissemination of content, as it is attractive and differentiated from other technologies and tools from the outset. The fact that the devices for its visualization have become simpler, cheaper, and more generalized makes it a different bet, especially suitable for young audiences and consumers of technology.

A review of the literature has revealed an increasingly frequent use of AR and its potential for the knowledge and dissemination of immovable cultural heritage. It is worth noting that the first papers published along these lines belonged to the field of computer sciences, but there are increasingly more articles from the social sciences and arts and humanities, which proves that these disciplines are also adding to the same trend. In the latest years, the uses of AR for the dissemination of cultural heritage have branched out to include education, exhibition improvement, exploration, reconstruction, and virtual museums.

Based on the revisions made, as well as on the analysis of the experiences related to the correct management of heritage, a decalogue of good practices has been drawn up in this paper. We believe that this decalogue should be followed when designing projects for the dissemination of architectural heritage that include these types of experiences and when planning recovery strategies in rural areas to ensure their success. In these cases, more than in any other, the scientific and technological components of an AR experience are subjects to social and economic aspects. It is about revitalizing these rural areas and attracting and establishing a young population through new initiatives that promote new possibilities for work and social wellbeing.

5. Conclusions

AR is becoming a very relevant tool in the dissemination of cultural heritage which, if used properly, can also help the development of rural areas and the maintenance of their vernacular architecture.

In this work we have presented a bibliometric analysis of the publications made in this field in recent years and an in-depth review of the literature. This revision allows us to conclude that, although it is clear that AR is a powerful tool for the dissemination of vernacular architecture, there are still no guidelines that define how to proceed when developing an AR application for this purpose. To fill this gap, this work proposes a decalogue of good practices that should inspire the creation of AR experiences to be beneficial for the socioeconomic environment in which they are developed. Points 1 to 5 of the decalogue are to be undertaken prior to the development of the AR application, and are dedicated to the documentation of the remains and the study of economic factors and tourism feasibility. The other five points (6 to 10) focus on the implementation of the application itself.

As an example, a proof of concept of an AR application for the visualization of the archaeological site "La Matilla" is described. The site was buried to continue the roadworks that led to its discovery, and currently the only evidence of its existence that survives are the 3D models generated by the authors of this document. These 3D models have been used to develop the proof of concept.

Author Contributions: Conceptualization, M.J.M. and P.M.; methodology, P.M. and M.J.M.; software, E.P.; validation, E.P.; investigation, M.J.M.; writing—original draft preparation, M.J.M. and E.P.; writing—review and editing, P.M.; visualization, E.P.; supervision, M.J.M. and P.M.; project administration, P.M.; funding acquisition, P.M. All authors have read and agreed to the published version of the manuscript.

Funding: This research was funded by Ministry of Science, Innovation and Universities of Spain, through the project "3D massive data processing. Segmentation, recognition and geometric modelling of non-conventional buildings", PID2019-108271RB-C32.

Conflicts of Interest: The authors declare no conflict of interest.

References

1. Edwards, S. Vernacular Architecture and the 21st Century ArchDaily. Available online: https://www.archdaily.com/155224/vernacular-architecture-and-the-21st-century/ (accessed on 17 March 2020).
2. Pérez Gil, J. Qué es la Arquitectura Vernácula? In *Historia y Concepto de un Patrimonio Cultural Específico*; Universidad de Valladolid: Valladolid, Spain, 2016; ISBN 978-84-8448-862-0.
3. Calderón Torres, C.M. *Arquitectura Vernácula de Extremadura I: Diseño de un Inventario*; Junta de Extremadura: Merida, Spain, 2011; ISBN 978-84-9852-309-6.
4. Bueno, G.; Granic, A.; Merchán, M.J.; Zaharija, G. Good practices in Heritage Buildings' Management. In *Innovation in Intelligent Management of Heritage Buildings (I2MHB)*; University of Zagreb: Zagreb, Croatia, 2019; pp. 29–40, ISBN 978-953-8250-04-0.
5. Cozzani, G.; Pozzi, F.; Dagnino, F.M.; Katos, A.V.; Katsouli, E.F. Innovative technologies for intangible cultural heritage education and preservation: The case of i-Treasures. *Pers. Ubiquitous Comput.* **2017**, *21*, 253–265. [CrossRef]
6. Bekele, M.K.; Pierdicca, R.; Frontoni, E.; Malinverni, E.S.; Gain, J. A Survey of Augmented, Virtual, and Mixed Reality for Cultural Heritage. *J. Comput. Cult. Herit.* **2018**, *11*, 1–36. [CrossRef]
7. Martins, J.; Sylaiou, S.; Bueno, G. COST Action TD1406—i2MHB. In *Innovation in Intelligent Management of Heritage Buildings (I2MHB)*; University of Zagreb: Zagreb, Croatia, 2019; pp. 11–19, ISBN 978-953-8250-04-0.
8. Pérez, E.; Merchán, P.; Merchán, M.; Salamanca, S. Fusion of 3D digitization technologies for the virtual exploration of re-covered archaeological remains. In Proceedings of the Science and Digital Technology for Cultural Heritage, Seville, Spain, 26–30 March 2019.
9. Carmigniani, J.; Furht, B.; Anisetti, M.; Ceravolo, P.; Damiani, E.; Ivkovic, M. Augmented reality technologies, systems and applications. *Multimed. Tools Appl.* **2011**, *51*, 341–377. [CrossRef]
10. Gleue, T.; Dähne, P. Design and implementation of a mobile device for outdoor Augmented Reality in the Archeoguide pro-ject. In Proceedings of the 2001 Conference on Virtual Reality, Archeology, and Cultural Heritage, Glyfada, Greece, 28–30 November 2001.
11. Vlahakis, V.; Karigiannis, J.; Tsotros, M.; Gounaris, M.; Almeida, L.; Stricker, D.; Gleue, T.; Christou, I.T.; Ioannidis, N. Io-annidis, Archeoguide first results of an augmented reality, mobile computing system in cultural heritage sites. In Proceedings of the 2001 Conference on Virtual Reality, Archaeology, and Cultural Heritage, Glyfada, Greece, 28–30 November 2001.
12. Fritz, F.; Susperregui, A.; Linaza, M. Enhancing cultural tourism experiences with augmented reality technologies. In Proceedings of the 6th International Symposium on Virtual Reality, Archaeology and Cultural Heritage, Pisa, Italy, 8–11 November 2005.
13. Kopp, S.; Gesellensetter, L.; Kramer, N.C.; Wachsmuth, I. "A Conversational Agent as Museum Guide—Design and Eval-uation of a Real-World Application. In Proceedings of the 2005 International Workshop on Intelligent Virtual Agents, Pisa, Italy, 8–11 November 2005.
14. Abate, A.F.; Acampora, G.G.; Ricciardi, S. An interactive virtual guide for the AR based visit of archaeological sites. *J. Vis. Lang. Comput.* **2011**, *22*, 415–425. [CrossRef]
15. Bickmore, T.; Pfeifer, L.; Schulman, D. Relational Agents Improve Engagement and Learning in Science Museum Visi-tors. In Proceedings of the 10th International Conference on Intelligent Virtual Agents, Philadelphia, PA, USA, 20–22 September 2011.
16. Calisi, D.; Tommasetti, A.; Topputo, R. Architectural historical heritage: A tridimensional multilayers cataloguing method. *ISPRS Int. Arch. Photogramm. Remote Sens. Spat. Inf. Sci.* **2012**, *38*, 599–606. [CrossRef]
17. Meschini, A.; Rossi, D.; Feriozzi, R. Disclosing Documentary Archives: AR interfaces to recall missing urban scenery. *Digit. Herit. Int. Congr.* **2013**, *2*, 371–374. [CrossRef]
18. Durand, E.; Merienne, F.; Pere, C.; Callet, P. Ray-on, an On-Site Photometric Augmented Reality Device. *J. Comput. Cult. Heritage* **2014**, *7*, 1–13. [CrossRef]
19. Rossi, D.; Petrucci, E.; Fazzini, S. A framework to increase the video-mapping accuracy of an architectural heritage mock-up. In Proceedings of the 2014 Virtual Reality International Conference, Laval, France, 9–11 April 2014.
20. Cordido, M.D.; Castañeda, O.M.; Soto, A.N.; Jiménez, W.B. The altarpieces of de la Candelaria in Guarenas and del Rosario Chapels in San Jacinto de Caracas Church: A graphic method to establish their historic relation. *Revista* **2016**, *180*, 38.
21. Cárdenas, M.I.Z.; De Medellín, U. Aplicación en realidad aumentada para divulgación del patrimonio cultural. *Kepes* **2016**, *13*, 33–59. [CrossRef]

22. Azmin, A.K.; Kassim, M.H.; Abdullah, F.; Sanusi, A.N.Z. Architectural heritage restoration of rumah datuk setia via mobile augmented reality restoration. *Plan. Malays. J.* **2017**, *15*, 139. [CrossRef]
23. Petrucci, E.; Rossi, D. Musepick: An integrated technological framework TO present the complex of santissima annunziata in ascoli piceno (Italy). *ISPRS Int. Arch. Photogramm. Remote Sens. Spat. Inf. Sci.* **2017**, *42*, 557–564. [CrossRef]
24. Albourae, A.T.; Armenakis, C.; Kyan, M. Architectural heritage visualization using interactive technologies. *ISPRS Int. Arch. Photogramm. Remote Sens. Spat. Inf. Sci.* **2017**, *42*, 7–13. [CrossRef]
25. Ioannidi, A.; Gavalas, D.; Kasapakis, V. Flaneur: Augmented exploration of the architectural urbanscape. In Proceedings of the 2017 IEEE Symposium on Computers and Communications (ISCC), Heraklion, Greece, 3–6 July 2017.
26. Morganti, C.; Bartolomei, C. Habanapp: Havana's architectural heritage a click away. *ISPRS Int. Arch. Photogramm. Remote Sens. Spat. Inf. Sci.* **2018**, *42*, 723–730. [CrossRef]
27. Banfi, F.; Brumana, R.; Stanga, C. Extended reality and informative models for the architectural heritage: From scan-to-BIM process to virtual and augmented reality. *Virtual Archaeol. Rev.* **2019**, *10*, 14–30. [CrossRef]
28. Germanese, D.; Pascali, M.A.; Berton, A.; Leone, G.R.; Moroni, D.; Jalil, B.; Tampucci, M.; Benassi, A. Architectural heritage: 3D documentation and structural monitoring using UAV. *CEUR Workshop Proc.* **2019**, *2320*, 1–12.
29. Shabalina, O.; Timofeev, G.; Davtian, A.; Zheltukhina, M.; Moffat, D. Investigating Regional Heritage through the Development and Playing of AR Games. In Proceedings of the European Conference on Games Based Learning, Hoff, UK, 23–24 September 2021.
30. Luigini, A.; Brusaporci, S.; Vattano, S.; Tata, A. 3D digital models for a widespread museum: The Renon's "bauernhöfe". *ISPRS Int. Arch. Photogramm. Remote Sens. Spat. Inf. Sci.* **2019**, *42*, 447–453. [CrossRef]
31. Matini, M.R.; Andaroodi, E.; Ono, K. A 3D approach to reconstitution of the adobe citadel of Bam after earthquake: A complementary interpretation of architectural heritage knowledge, aerial photogrammetry, and heterogeneous data. *Int. J. Arch. Heritage* **2018**, *13*, 600–618. [CrossRef]
32. Rashid, M.; Antlej, K. Geospatial platforms and immersive tools for social cohesion: The 4D narrative of architecture of Australia's Afghan cameleers. *Virtual Archaeol. Rev.* **2020**, *11*, 74–84. [CrossRef]
33. Li, S.-S.; Tai, N.-C. Development of an Augmented Reality Application for Protecting the Perspectives and Views of Architectural Heritages. In Proceedings of the 2020 IEEE International Conference on Consumer Electronics Taiwan, Taoyuan, Taiwan, 28–30 September 2020.
34. Lin, K.-H.; Tai, N.-C. In-Depth Learning of Architectural Heritage with Application of Augmented Reality based on Sequential Scenes. In Proceedings of the 2020 IEEE International Conference on Consumer Electronics Taiwan, Taoyuan, Taiwan, 28–30 September 2020.
35. Templin, T.; Popielarczyk, D. The Use of Low-Cost Unmanned Aerial Vehicles in the Process of Building Models for Cultural Tourism, 3D Web and Augmented/Mixed Reality Applications. *Sensors* **2020**, *20*, 5457. [CrossRef]
36. Aicardi, I.; Chiabrando, F.; Lingua, A.M.; Noardo, F. Recent trends in cultural heritage 3D survey: The photogrammetric computer vision approach. *J. Cult. Heritage* **2018**, *32*, 257–266. [CrossRef]
37. Spring, A.P.; Peters, C.; Minns, T. Using Mid-Range Laser Scanners to Digitize Cultural-Heritage Sites. *IEEE Eng. Med. Biol. Mag.* **2010**, *30*, 15–19. [CrossRef]
38. Yu, D.; Jin, J.S.; Luo, S.; Lai, W.; Huang, Q. A Useful Visualization Technique: A Literature Review for Augmented Reality and its Application, Limitation & Future Direction. In *Visual Information Communication*; Springer International Publishing: Berlin, Germany, 2009; pp. 311–337.
39. Faion, F.; Zea, A.; Noack, B.; Steinbring, J.; Hanebeck, U.D. Camera- and IMU-based pose tracking for augmented reality. In Proceedings of the 2016 IEEE International Conference on Multisensor Fusion and Integration for Intelligent Systems (MFI); IEEE: Berlin, Germany, 2017; pp. 648–653.
40. Gustafsson, F.; Schön, T.B.; Hol, J.D. Sensor Fusion for Augmented Reality. In *IFAC Proceedings Volumes*; Elsevier BV: Amsterdam, The Netherlands, 2008; Volume 41, p. 14100.
41. Bellotti, V.; Price, B.; Rasmussen, P.; Roberts, M.; Schiano, D.J.; Walendowski, A.; BeGole, B.; Chi, E.H.; Ducheneaut, N.; Fang, J.; et al. Activity-based serendipitous recommendations with the Magitti mobile leisure guide. In Proceedings of the Twenty-Sixth Annual CHI Conference on Human Factors in Computing Systems, Florence, Italy, 5–10 April 2008.
42. Burke, J.; Estrin, D.; Hansen, M.; Parker, A.; Ramanathan, N.; Reddy, S.; Srivastava, M. Participatory Sensing. In Proceedings of the World Sensor Web Workshop, Como, Italy, 19–20 September 2013.
43. Keating, G.; Guest, D.; Konertz, A.; Padovani, N.; Villa, A. Designing the AR Experience: Tools and Tips for Mobile Augmented Reality UX Design. In *Constructive Side-Channel Analysis and Secure Design*; Springer International Publishing: Berlin, Germany, 2011; Volume 6770, pp. 135–141.
44. Barfield, W.; Baird, K.M.; Bjorneseth, O.J. Presence in virtual environments as a function of type of input device and display update rate. *Displays* **1998**, *19*, 91–98. [CrossRef]
45. Klopfer, E. Augmented Learning. *Augment. Learn.* **2008**. [CrossRef]
46. Hartson, H.R. Cognitive, physical, sensory, and functional affordances in interaction design. *Behav. Inf. Technol.* **2003**, *22*, 315–338. [CrossRef]
47. Lee, W.-H.; Lee, H.-K. The usability attributes and evaluation measurements of mobile media AR (augmented reality). *Cogent Arts Humanit.* **2016**, *3*, 1–16. [CrossRef]

48. Anderson, R.H.; Shapiro, N.Z. *Beyond User Friendly;* Rand Corp: Santa Monica, CA, USA, 1989; pp. 50–54.
49. Nielsen, J. *Usability Engineering;* Morgan Kaufmann: Burlington, MA, USA, 1993; ISBN 978-0125184069.
50. Dünser, A.; Hornecker, E.; Duenser, A. Lessons from an AR book study. In Proceedings of the 1st International Conference on Medical and Health Informatics, Taiwan, China, 20–22 May 2017.
51. Cameron, F.; Kenderdine, S. *Theorizing Digital Cultural Heritage: A Critical Discourse;* The MIT Press: Cambridge, MA, USA, 2007; ISBN 978-0-262-03353-4.
52. Breuss-Schneeweis, P. The speaking celt. In Proceedings of the 2016 ACM International Joint Conference on Pervasive and Ubiquitous Computing, Heidelberg, Germany, 12–16 September 2016.
53. Miyashita, T.; Meier, P.; Tachikawa, T.; Orlic, S.; Eble, T.; Scholz, V.; Gapel, A.; Gerl, O.; Arnaudov, S.; Lieberknecht, S. An Augmented Reality museum guide. In Proceedings of the 2008 7th IEEE/ACM International Symposium on Mixed and Augmented Reality, Cambridge, UK, 15–18 September 2008.
54. Skjermo, J.; Stokes, M.J.; Hallgren, T.; Kofod-Petersen, A. Towards a Simple Augmented Reality Museum Guide. In Proceedings of the Norwegian Artificial Intelligence Symposium, Trondheim, Norway, 27–28 May 2019.
55. Traum, D.; Aggarwal, P.; Artstein, R.; Foutz, S.; Gerten, J.; Katsamanis, A.; Leuski, A.; Noren, D.; Swartout, W. Ada and Grace: Direct Interaction with Museum Visitors. In *Constructive Side-Channel Analysis and Secure Design;* Springer International Publishing: Berlin, Germany, 2012; pp. 245–251.
56. Lane, H.C.; Cahill, C.; Foutz, S.; Auerbach, D.; Noren, D.; Lussenhop, C.; Swartout, W. The Effects of a Pedagogical Agent for Informal Science Education on Learner Behaviors and Self-efficacy. In *Constructive Side-Channel Analysis and Secure Design;* Springer International Publishing: Berlin, Germany, 2013; pp. 309–318.
57. Vosinakis, S.; Avradinis, N. Virtual agora: Representation of an ancient greek agora in virtual worlds using biological-ly-inspired motivational agents. *Mediterr. Archaeol. Archaeom.* **2016**, *16*, 29. [CrossRef]
58. Gimeno, J.; Portalés, C.; Coma, I.; Fernández, M.; Martínez, B. Combining traditional and indirect augmented reality for indoor crowded environments. A case study on the Casa Batlló museum. *Comput. Graph.* **2017**, *69*, 92–103. [CrossRef]
59. Santachiara, M.; Gherardini, F.; Leali, F. An Augmented Reality Application for the Visualization and the Pattern Analysis of a Roman Mosaic. *IOP Conf. Series: Mater. Sci. Eng.* **2018**, *364*, 012094. [CrossRef]
60. Siekański, P.; Bunsch, E.; Sitnik, R. Seeing the past: An augmented reality application for visualization the previous state of cultural heritage locations. *Electron. Imaging* **2018**, *2018*, 452. [CrossRef]
61. Rowe, J.P.; Lobene, E.V.; Mott, B.W.; Lester, J.C. Play in the Museum. *Int. J. Gaming Comput. Simul.* **2017**, *9*, 96–113. [CrossRef]
62. Younes, G.; Kahil, R.; Jallad, M.; Asmar, D.; Elhajj, I.; Turkiyyah, G.; Al-Harithy, H. Virtual and augmented reality for rich interaction with cultural heritage sites: A case study from the Roman Theater at Byblos. *Digit. Appl. Archaeol. Cult. Herit.* **2017**, *5*, 1–9. [CrossRef]
63. Grijalba Bengoetxea, J.; Grijalba Bengoetxea, A.; Kamal Chapagain, N.; Nair, P.K.V. *Best Practices Manual in Cultural Heritage Management;* Universidad de Valladolid: Valladolid, Spain, 2016; ISBN 978-81-933125-0-6.
64. Eppich, R.; Grinda, J.L.G. Management Documentation: Indicators & Good Practice at Cultural Heritage Places. *ISPRS Int. Arch. Photogramm. Remote Sens. Spat. Inf. Sci.* **2015**, *5*, 133–140. [CrossRef]
65. De Beer, J.; Boogaard, F. Good practices in cultural heritage management and the use of subsurface knowledge in urban areas. *Procedia Eng.* **2017**, *209*, 34–41. [CrossRef]
66. Guzmán, P.; Roders, A.A.P.; Colenbrander, B.B. Measuring links between cultural heritage management and sustainable urban development: An overview of global monitoring tools. *Cities* **2017**, *60*, 192–201. [CrossRef]

applied sciences

MDPI

Article

Design and Evaluation of a Web- and Mobile-Based Binaural Audio Platform for Cultural Heritage

Marco Comunità *,†, **Andrea Gerino, Veranika Lim and Lorenzo Picinali** *,†

Dyson School of Design Engineering, Imperial College London, London SW7 2DB, UK; a.gerino15@imperial.ac.uk (A.G.), v.lim@imperial.ac.uk (V.L.)

* Correspondence: m.comunita@imperial.ac.uk (M.C.), l.picinali@imperial.ac.uk (L.P.)

† These authors contributed equally to this work.

Abstract: PlugSonic is a suite of web- and mobile-based applications for the curation and experience of 3D interactive soundscapes and sonic narratives in the cultural heritage context. It was developed as part of the PLUGGY EU project (Pluggable Social Platform for Heritage Awareness and Participation) and consists of two main applications: PlugSonic Sample, to edit and apply audio effects, and PlugSonic Soundscape, to create and experience 3D soundscapes for headphones playback. The audio processing within PlugSonic is based on the Web Audio API and the 3D Tune-In Toolkit, while the mobile exploration of soundscapes in a physical space is obtained using Apple's ARKit. The main goal of PlugSonic is technology democratisation; PlugSonic users—whether cultural institutions or citizens—are all given the instruments needed to create, process and experience 3D soundscapes and sonic narratives; without the need for specific devices, external tools (software and/or hardware), specialised knowledge or custom development. The aims of this paper are to present the design and development choices, the user involvement processes as well as a final evaluation conducted with inexperienced users on three tasks (creation, curation and experience), demonstrating how PlugSonic is indeed a simple, effective, yet powerful tool.

Keywords: binaural spatialisation; spatial audio; cultural heritage; digital heritage; sonic narratives; soundscapes; web audio; augmented reality

Citation: Comunità, M.; Gerino, A.; Lim, V.; Picinali, L. Design and Evaluation of a Web- and Mobile-Based Binaural Audio Platform for Cultural Heritage. *Appl. Sci.* **2021**, *11*, 1540. https://doi.org/10.3390/app11041540

Academic Editor: Marek Milosz
Received: 14 January 2021
Accepted: 2 February 2021
Published: 8 February 2021

1. Introduction

A heritage that is everywhere, and is relevant to everyday life, is one of the preconditions for genuine sustainability [1]. Currently, there are few Information and Communication Technologies (ICT) tools to support citizens in their everyday activities to shape cultural heritage and be shaped by it. Furthermore, existing applications and repositories for heritage dissemination do not foster the creation of heritage communities (e.g., Google Arts and Culture, and Europeana) [2,3]. Social platforms certainly offer potential to build networks, but they have not yet been fully exploited for global cultural heritage promotion and integration in people's everyday life [4] and museums and institutions have only recently started to explore the potential of social media and technology for public engagement and co-creation purposes [5]. The PLUGGY project (Pluggable Social Platform for Heritage Awareness and Participation) [6] aims to bridge this gap by providing the necessary tools to allow users to share their local knowledge and everyday experience with others and, together with the contribution of cultural institutions, to build extensive networks around a common area of interest, connecting the past, the present and the future.

Within PLUGGY, several tools are being developed (Figure 1): a social platform, curation tools and four "pluggable" apps to demonstrate the platform's potential and kick-start applications for the after-project life. The social platform serves both as a repository for all content uploaded (assets) and curated (exhibitions), as well as a place for interaction and collaboration. It has all the typical features of a social platform: profile, private and public content; follow; like; comment; share; notifications; folders to organise bookmarked content;

and teams for collective contribution. At the same time, the social platform gives access to the curation tools—which are designed to create basic exhibitions in the form of blog stories or time-lines—and is extended by the "pluggable" apps for the curation of augmented exhibitions and engaging experiences. These applications focus on various aspects of digital heritage, which include Virtual (VR) and Augmented Reality (AR), Geolocation, Gamification and 3D Soundscapes and Sonic Narratives. The latter, called PlugSonic, is the focus of the current paper, which will look in particular at the web- and mobile-based binaural audio features of the application. The PLUGGY social platform gives also access to external repositories like Europeana [7] and Wikipedia and includes an open source application programming interface (API) which allows anyone to develop new applications that access the platform's content and extend its features or curation tools. Any user can request to become a developer and—after adequate evaluation and approval—"plug" a new app to the platform for everyone to benefit (Figure 1).

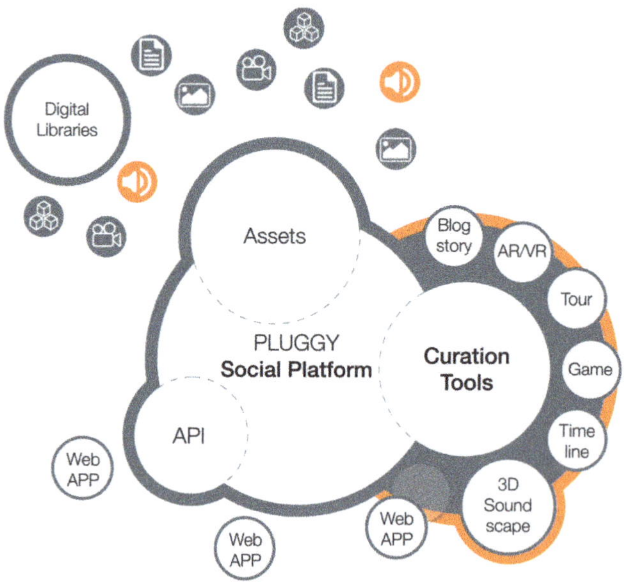

Figure 1. The PLUGGY social platform and apps structure.

The research presented here aims to contribute to both the cultural heritage and spatial audio communities through the development of novel tools for the creation and experience of realistic and interactive 3D soundscapes. We describe the design choices and the key implementation details and we also illustrate the authoring workflow. We show how 3D audio and augmented reality technologies can be exploited to enhance visits to museums and monuments [8]. These could also be adopted to document and preserve the acoustics of an archaeological site [9] or landscape [10] or to document acoustic heritage [11]. By integrating web- and mobile-based applications with a social platform and online repository we show how binaural audio and sonic narratives could be used to encourage cultural heritage dissemination and availability.

In this work, special importance has been given to designing applications which allow users with no previous knowledge or experience in 3D audio production or soundscapes curation, to contribute to the popularisation and growth of sonic cultural heritage. Being web- and mobile-based, the tools presented here are intrinsically ubiquitous and the content can be experienced online or in a physical space, without the need for specialised software or hardware installation and/or custom development. To understand whether these goals were achieved we conducted an evaluation with inexperienced users to substantiate our objective towards a very low barrier to the content development, while retaining the effectiveness in public engagement of spatial audio technologies. In this respect, focus of our evaluation was the usability of our tools, especially for naive users (no expert sound designers were involved). Therefore, quality of the output, pleasantness, accuracy and realism of the simulation are not within the scope of this work. There are already many tools available for expert users (see Section 2) and studies around the trade-offs between accuracy and plausibility of a sonic interaction [12].

The paper is organised as follows. Section 2 provides a background about binaural spatialisation, as well as a review of the state of the art for web-based spatial audio and the use of sonic narratives in cultural heritage. Section 3 lists the research aims. Section 4 describes all the apps that make up the *PlugSonic Suite*, main functionalities (complete details are included in the appendices) and implementation choices. In Section 5, we report on the evaluation methods and discuss the results. Section 6 is devoted to the conclusions and potential future work.

2. Background

2.1. Sonic Narratives and Web-Based Spatial Audio

A sonic narrative is defined as a sequence of sounds that suggests a sense of causality and temporal evolution. This is usually achieved through the interplay of tension and resolution in the acoustic properties of the sequence, or through its referential qualities [13]. Generally based on music features (e.g., timbre, pitch-melody, tempo, etc.) [14], sonic narratives are often not interactive (i.e., simple audio playback). Conversely, a soundscape is defined as the sum of natural, animal and human sounds that describe a landscape or environment [15] and—while dynamic at the single acoustic events level—it typically describes a static scene, with no control over the listening or sound sources' position and/or proximity.

The aim of binaural spatialisation is to provide the listener (through standard headphones) with the impression that sound is positioned in a specific location in the three-dimensional space. The 3D characteristics of the sound can be captured during recording with special hardware, or simulated in post-production via spatialisation techniques. The addition of spatial attributes (e.g., placement of sound sources on a full 360° sphere, and at different distances), and most of all the addition of interactivity (e.g., to navigate soundscapes or sonic narratives moving around in the acoustic virtual environment), are features which have not been widely explored until now within the context of digital heritage. For example, the authors of [16] explored spatial sonic narratives, but exploited simple 2-dimensional audio panning techniques.

The theories at the basis of the binaural spatialisation technique are not particularly recent, and the first binaural recording dates back to the end of the 19th Century [17]. However, it is only within the last twenty years that the increase in the calculation power of personal computers enabled an accurate real-time simulation of three-dimensional sound-fields over headphones.

With the release of the Web Audio (WAA) [18] and the Web GL (WGL) [19] APIs in 2011, and with the specification of HTML5 in 2014, the World Wide Web Consortium (W3C) and Mozilla Foundation set the basis for the development of modern web applications. As stated in the introduction to the WAA [20], the specification of a high-level Javascript (JS) API was necessary to satisfy the demand for audio and video processing capabilities that would allow to develop "sophisticated web-based games or interactive applications". To obtain performances comparable to modern digital audio workstations (DAWs) and games graphical engines, the two organisations decided to move the burden of audio and video processing from the server to the client (i.e., browser) side.

Several tools currently exist that perform binaural spatialisation, just to mention a few: Anaglyph [21], IRCAM Spat [22], the IEM binaural audio open library [23] and the 3D Tune-In Toolkit [24]. Although, very few of these are designed for non-expert users or implemented within a web-based application, and therefore available on multiple platforms through a simple browser. The idea of developing and evaluating web and mobile applications for the creation and experience of 3D sonic narratives and soundscapes does indeed represent a novel contribution to both the digital heritage and audio technology domains.

2.2. Sonic Narratives for Cultural Heritage

In this section, we look at how concepts like sonic narratives, soundscapes and interactivity have been explored in the cultural heritage context. We also look at audio and/or web technologies adoption and implementation methods, the results achieved so far, and the limitations of current systems, when it comes to cultural heritage applications—describing the panorama in which our system is being proposed.

In [25], Ardissono et al. give an exhaustive review and comparative analysis of research about digital storytelling and the delivery of multi-media content, on-site or on-line, with a focus on cultural heritage. Here, we will limit to those projects that use exclusively or mainly audio as medium for content delivery or to design novel types of experiences.

Looking at the state of the art in this area, it can be noticed how research is delving into solutions to make cultural heritage immersive, adopting augmented reality (AR), virtual reality (VR) and spatial audio; engaging, using personalisation and emotional storytelling; adaptive, exploiting context-awareness and location-awareness; interactive, using the paradigm of dramas; or open and inclusive, developing content for people with impairments and/or difficulties.

Two of the first projects aiming at enriching the user experience and engagement in a museum's visit were the HyperAudio and the HIPS projects. In [26], the authors introduce the HyperAudio project, in which the user is presented with audio content played through headphones when approaching an artefact, together with suggestions for further exploration on the display of a palmtop. In [27], the hardware and software architecture is presented, which uses infrared transmitters nearby the exhibits and a receiver on the user's headphones to implement a degree of location awareness. The HIPS project [28] expands on the HyperAudio experience aiming at a user-adaptive presentation system.

The LISTEN project [29] investigated audio augmented environments and user adaptation technologies. This involved the development of ListenSpace [30]—graphical authoring tool used to represent the real space and the sound sources' position—as well as the implementation of a domain ontology [31] for an exhibition and the use of context-awareness to adapt to the user's interest. For the spatial rendering the authors relied on IRCAM's Spatialisateur [32]. The main limitations, from the content creators' perspective, could be seen into the system's software (server-based processing) and hardware (antennas or infrared cameras for head-tracking) requirements, and the necessity for custom development for each exhibition, together with the sound rendering only for the horizontal plane.

In the CHESS project [33], the focus was on personalisation and interactivity; where the story was delivered mainly through voice narration, with on-screen instructions and interaction obtained through applications running on web browsers. Personalisation was obtained through the definition of personae, profiling first-time visitors [34] to adapt the narration style. Once profiled, the adaptive storytelling engine [33] used contextual data to adapt the user's experience.

Interaction and context awareness (based on geolocation) was explored in [35] with mobile urban dramas (in which the user becomes the main character of a story). The project used a multimedia style (audio, video, images and animations) and was implemented to run on mobile web browsers using XML to describe the content. Here, the advantage of multi-platform flexibility, was limited by the need for specific knowledge about the content metadata structure or the consultancy from the researchers for app implementation and web services.

Other projects have been studying and developing new ways to attract visitors with engaging experiences. The EMOTIVE project [36] aimed to "research, design, develop and evaluate methods and tools that can support the cultural and creative industries in creating narratives" that exploit emotional storytelling. A first evaluation of a mobile application for the ancient Athens Agora [37] produced positive results with the users particularly appreciating the chance of freely exploring the environment.

In the ARCHES project [38], researchers worked on the inclusivity aspect of storytelling exploring new modalities to design cultural heritage experiences for people with difficulties and/or disabilities. Other examples of tools developed by academia that could be used to create soundscapes and immersive experiences are the Soundscape Renderer [39] and those developed for the CARROUSO project [40].

There are also commercial applications developed for audio narratives like Echoes.XYZ [41]; both are GPS-triggered audio tours with web-based tools for the creation of audio guides and mobile apps to experience them. It is worth highlighting how none of the currently available commercial solutions use sound spatialisation.

3. Aims

The overall aims of our research are illustrated below.

- Design and develop tools that foster spatial audio for the creation of interactive soundscapes and sonic narratives, with a focus towards cultural heritage.
- Adopt web and mobile technologies to simplify and streamline the curation process and prevent the need for specialised software, custom applications development and/or hardware requirements.
- Prove—through an evaluation with inexperienced participants—that 3D audio technologies can be designed to be accessible while remaining capable of delivering an engaging and compelling experience.

4. PlugSonic Suite

In this section, we describe the design criteria and some implementation details of the web and mobile applications that make up the *PlugSonic Suite* (for complete implementation details, see in [42]). The apps—organised into the PlugSonic Sample and PlugSonic Soundscape groups—were developed to (1) facilitate the use of audio content to augment virtual exhibitions; (2) enhance on-line and/or on-site visits to museums, monuments or archaeological sites; and (3) share tangible and intangible cultural heritage.

Specifically, we designed PlugSonic Sample to edit sound files and apply audio effects, and PlugSonic Soundscape to create and experience interactive spatialised soundscapes.

In this way, social platform, curation tools and pluggable apps can include standard sound files (i.e., mono/stereo or mp3/wav format) to be used in voice descriptions, audio narratives or sound accompaniment to the platform's exhibitions, as well as interactive and explorable 3D audio narratives and soundscapes. PLUGGY users—whether institutions or citizens—are therefore given all the necessary instruments, and are not in need for specific devices, external tools (software and/or hardware), specialised knowledge and/or resources.

Figure 2 gives a complete overview of the structure of the web and mobile apps included in the *PlugSonic Suite*. PlugSonic Sample and PlugSonic Soundscape Create were implemented only as web applications while PlugSonic Soundscape Experience was implemented both as web and mobile application. In this way, all the Soundscape exhibitions in PLUGGY can be explored in a virtual or physical space. When using Soundscape Experience Web, the navigation takes place on the browser using mouse, arrow keys or touch controls. With Soundscape Experience Mobile the navigation can take place in a real space; thanks to the use of the Apple ARKit [43], the user can freely navigate within a room while the device's camera is used to extrapolate the person's position in the room and hence in the soundscape.

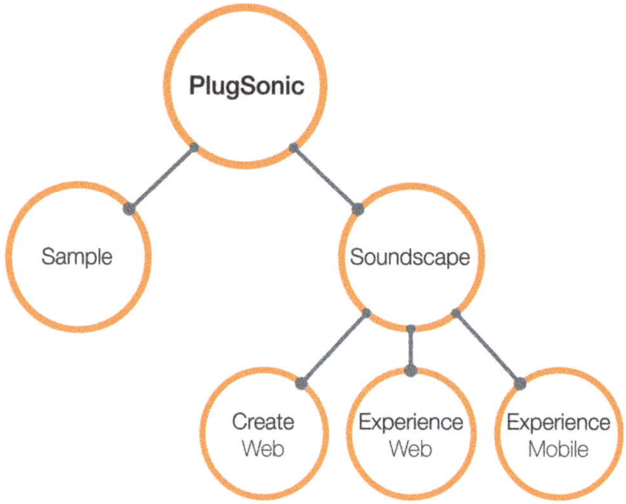

Figure 2. Apps of the *PlugSonic Suite*.

A typical use case of the *PlugSonic Suite* could be described as follows. Let us suppose that a user would like to curate a new 3D audio soundscape. The first step would be to select the appropriate audio assets to be included in the Soundscape exhibition. The user has the option to either upload new audio files to the PLUGGY social platform or use content already available. Any file uploaded to PLUGGY (e.g., image, video, 3D model or sound) is associated to an asset—with a dedicated web page—which contains extended information such as title, description, location, license, tags, number of likes, comments, etc. While creating a new audio asset, the user might decide to use PlugSonic Sample to edit the audio file and enhance the sound. The next step would be to create the new Soundscape exhibition using PlugSonic Soundscape Create. Upon loading the app, the user can set size and shape of the virtual environment (which might relate to a real physical space or not) and search and retrieve the desired audio files. The sound sources are imported into the soundscape and represented as *circles* in the virtual room. The curation process includes setting the position for each sound source together with many other options controlling the sources' sound and the interaction between sources and user (see Section 4.2 for details). As PlugSonic Soundscape Create allows for playback and navigation, at any moment during the curation, the user can listen to the soundscape and test the exhibition. After this curation steps are completed, the user can save and publish it. At this point, other users could explore the soundscape using their own device (personal computer, mobile or tablet). Some might choose to do so from home, using Soundscape Experience Web to visit the exhibition in the virtual environment. Others might be in the real space (e.g., gallery, museum or archaeological site), or might just decide to explore the soundscape through physical movements within a space, and would therefore choose the Soundscape Experience Mobile app.

During the design process—and whenever a decision needed to be taken about user experience/interface, accessibility and ease of use—the authors' intent was to follow the principle stated in the Article 4 of the Convention on the Value of Cultural Heritage for Society—Faro Convention, 2005 [44]: "Everyone, alone or collectively, has the right to benefit from the cultural heritage and to contribute towards its enrichment." Therefore, differently from most of the tools described in previous sections, the focus of our project on inclusivity and participation required us to develop intuitive and immediate tools, usable by anyone without specific training, and ensuring a true impact on cultural heritage.

4.1. PlugSonic Sample

As explained in the introduction, PlugSonic Sample can be used to edit and apply audio effects to any audio file uploaded to the PLUGGY social platform. The modified file can then be saved into the social platform and used in any exhibition (e.g., Blog Story, Time Line and Soundscape).

Figure A1 shows the UI of the web application, integrated into the PLUGGY social platform. The UI is divided into three main parts: (1–2) waveform display with mouse navigation and selection functions; (3–14) playback and edit controls, together with buttons to export the audio file and save the modifications; and (15–18) filters and effects menu.

For a complete description of the controls and features implemented in Sample we refer to the Appendix A.

4.2. PlugSonic Soundscape Create

As illustrated in the previous sections, PlugSonic Soundscape Create was developed for the curation of interactive 3D audio narratives and soundscapes. To do so, a user proceeds with the creation of a new exhibition through the PLUGGY social platform. After selecting *Soundscape* as the exhibition's type and setting its title and description the user is presented with the Soundscape Create UI, allowing them to proceed with the curation.

The UI, shown in Figure A2, is divided into three main sections: (1–3) the "room", which shows the physical/virtual space described in the soundscape and includes icons that represent sound sources and listener; (4–7) the top bar, hosting playback control buttons; and (8–12) the dismissable side menu, containing all the controls and options to modify the soundscape.

The curation process would typically proceed as follows. The user starts by setting shape (round/rectangular) and size (width/depth/height) of the room and, if desired, choose an image to be used as floor-plan (*room* tab—Figure 3C). After uploading the sound sources—using the *search* tab (Figure 3A)—the user sets the options for each sound source (*sound sources* tab—Figure 3B). Appendix B includes a complete description of all the options available for each sound source, here we limit our description to what we believe are the most interesting and useful controls: *Position*—absolute or relative to the listener; *Loop*—to choose if the sound source will loop or play only once; *Reach*—to control the interaction between listener and sound source (when ON the listener will be able to hear the sound source only when inside the interaction area); *Reach radius*—to set the size of the interaction area; and *Timings*—to set an order in the reproduction of any sound source by constraining the playback to the reproduction of another sound source.

At any point during the soundscape's creation, the user can explore the soundscape to verify the results. It is also possible to take a recording while navigating the soundscape in real-time and export it as a .wav audio file, which will include all the properties of the 3D audio rendering in a standard stereo audio file.

The exhibition's properties (Figure A2) can be modified if necessary and the soundscape saved to the social platform and published. There are also buttons available to export the soundscape's metadata. This is to allow the user to keep a local copy of the soundscape. Furthermore, the metadata can be imported into any application capable of interpreting it. In the case of PlugSonic, the metadata can be imported back into Soundscape Create or it can be opened with Soundscape Experience Mobile (see Section 4.4). There are two formats available to export the soundscape. The simple metadata, which requires an internet connection to access PLUGGY's social platform and retrieve the audio files, or the metadata including the assets, in which case, the soundscape can be experienced off-line as the audio files' data is added to the exported soundscape file.

The UI presented here is the result of both a complete redesign and extension, as well as the outcome of the expert's evaluation described in [45]. For a complete description of the controls and features available in Plugsonic Soundscape Create we refer to Appendix B.

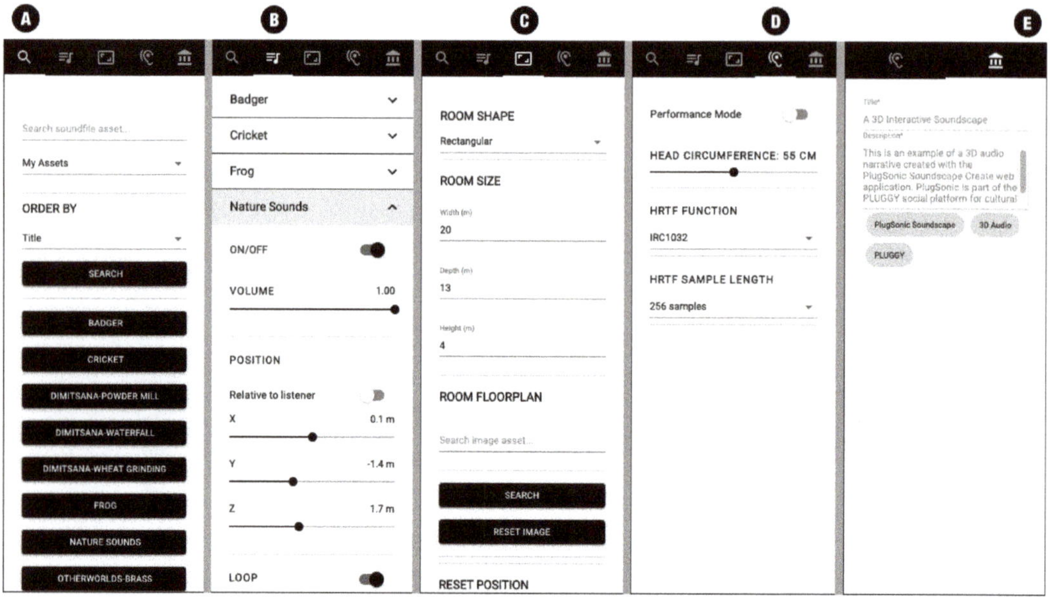

Figure 3. PlugSonic Soundscape Create dismissable menus: (**A**) Sound source search tab, (**B**) Sound sources settings tab, (**C**) Room settings tab, (**D**) Listener settings tab, (**E**) Exhibition tab

4.3. PlugSonic Soundscape Experience Web

The Soundscape Experience web app was developed to allow the navigation of *Soundscape* exhibitions using any device capable of running a web browser (pc, laptop, tablet or mobile). The app's UI is the same as Soundscape Create stripped down of all the features that allow to modify the soundscape—the only controls available are the playback and record buttons, the access to the touch arrows controls, and the listener's options (Figure 3D). Furthermore, the exhibition's title, description and tags are visible but not editable (Figure 3E). Soundscape Experience is loaded when a user clicks on the *view* button on the exhibition's page within the social platform.

4.4. PlugSonic Soundscape Experience Mobile

The Soundscape Experience Mobile application has been designed with two main goals: first, to enable navigating soundscapes using a natural, touch-based interface at home or on the go. Second, to allow users to explore soundscapes in an immersive virtual experience delivered in real-world environments. These goals are accomplished by providing two separate interaction modes.

In the first mode (Figure 4a), users can explore the soundscape by moving the listener's icon using their finger and can change orientation by rotating the device. The 3D audio simulation is updated in real-time according to the listener's icon current position.

The second interaction mode aims to provide an immersive experience by enabling users to explore a soundscape according to their movements in the real-world. In order to achieve such goal, we use ARkit [43], a technology developed by Apple to easily support detecting and tracking planar surfaces by analysing video frames captured by a device's camera and data collected by inertial sensors. The framework provides anchors in a real-world environment that can be used to determine the relative position of the device with respect to the detected plane. Such position is used to update the 3D audio simulation in real-time (Figure 4b).

Soundscapes can be loaded from the PLUGGY social platform using the QR code reader included in the app or importing the metadata. The app includes buttons to reset the listener's orientation, play and stop, and choose the HRTFs. It also allows some control over the convolutional reverb settings. The reverb uses impulse responses for three standard room sizes available from the 3D TuneIn Toolkit [24] but it could be extended by allowing to import other impulse responses.

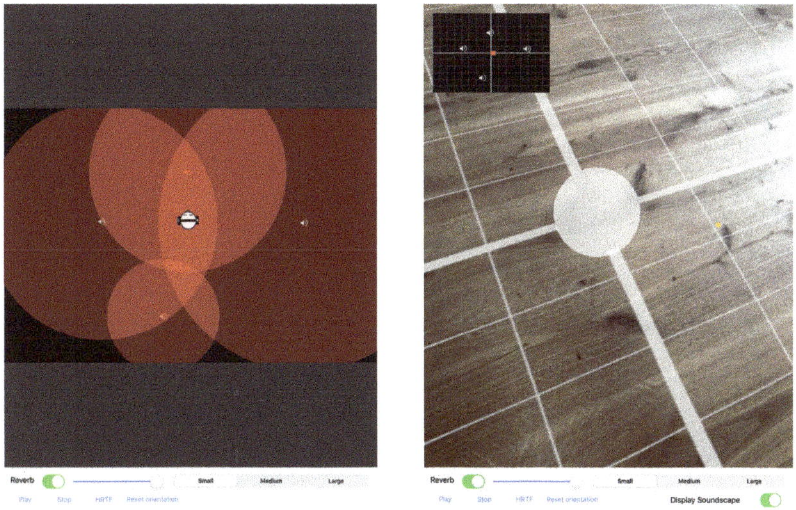

(**a**) Touch-based (**b**) Camera- and sensors-based

Figure 4. PlugSonic Soundscape Experience Mobile UI.

5. Evaluation

The evaluation of PlugSonic focused on the Soundscape Create and Soundscape Experience Mobile apps, which we considered the most critical in terms of contributions to both the spatial audio and cultural heritage research communities. The purpose of the evaluation was manifold: Understand whether users with no previous experience in sound design or cultural heritage content curation, and without previous knowledge and experience of 3D audio, could easily familiarise with the apps and use the functionalities and features as they were designed for. See if users could quickly and effectively use PlugSonic to recreate a 3D soundscape that maps onto a real physical space. Find out if spatial audio technologies in general and the Experience Mobile app in particular constitute a useful and practical way to design compelling experiences and improve engagement with and understanding of cultural heritage. Specifically, we intended to answer the following questions:

- How would users create a soundscape using PlugSonic?
- Could users easily find functionalities/features they need?
- Did users understand all functionalities/features offered in Plugsonic?
- If not, what issues did participants face?
- How long did it take them to successfully use the functionality/feature?
- How easy or difficult was it for users to recreate a 3D soundscape experienced in a real physical space?

The evaluation was arranged in three parts. In the first part, participants created a soundscape following predefined tasks provided to them. In the second part, the same participants listened to a soundscape set up in a real physical space and were then asked to recreate it using Soundscape Create. In the third part, another group of participants was asked to explore the soundscape created for part 2 using Soundscape Experience Mobile. We discuss the details further below.

5.1. Part 1—Soundscape Creation

5.1.1. Methodology

In the first part, participants were asked to create a soundscape for one of the institutions involved in the project, the Open-Air Water Power Museum in Dimitsana (Greece). They were asked to use the Soundscape Create web app together with the material (images and sounds) already available on the PLUGGY social platform. Initially, participants were given a verbal introduction to the PLUGGY project and shown the PLUGGY social platform website, how it is organised and how to navigate it. They were then given 5 min to get familiar with the application, explore the different features and ask questions for clarifications whenever necessary. Participants were then given 12 tasks (Table 1) that would lead to a complete Soundscape exhibition (Figure 5). Participants were asked to think aloud and, after each task, to rate how easy it was on a 7-point Likert scale—with 1 as *extremely difficult* and 7 as *extremely easy* (Single Ease Questions (SEQ)). Time required to complete each task was measured and is compared with baseline times from the first author. Observations were taken throughout the experiment. We recruited 5 participants for part 1: 2 males and 3 females, young academics in their mid 20s to mid 30s. All participants reported to have normal hearing and no previous experience with 3D soundscapes.

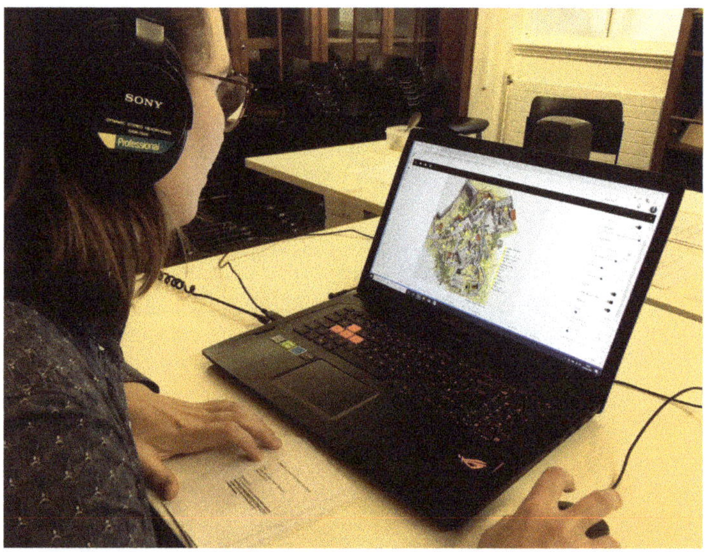

Figure 5. Participant using the Soundscape Create app for part 1 of the evaluation.

5.1.2. Results

Overall, most of the steps in task 1 were easy for participants to complete and the average rate was 6.4 (out of 7). See Table 1 for the average ease score and completion time per task.

However, participants faced some difficulties for 3 sub-tasks. For example, users had difficulties adding a new sound source. One participant did not use the *My Assets* option for the search. One participant exited the app to search in the "My Assets" page of the social platform and then tried the *Import* button meant to be used for Soundscape metadata (see Section 4.2). Once assets were found, 2 participants tried to drag and drop the sound source button from the search panel onto the room instead of clicking on it. Once an asset was added, one participant could not see the icon of the source as it was masked by the floor-plan image. Participants also had difficulties understanding the hinting system, as it was perceived as not giving the right suggestions based on the word being typed. Two participants could not set up the reach of the sound and 2 other participants had issues with the slider precision.

Table 1. Evaluation: Part 1—Single Ease Questions scores and completion times (average and standard deviation).

	Task Description	SEQ Avg (Std)	Completion Time (s) Avg (Std)	Baseline Completion Time (s)
1	CREATE a new Soundscape exhibition	6.8 (0.5)	73 (35.9)	66
2	Set the ROOM SIZE	7.0 (0.0)	16 (5.9)	10
3	Set the ROOM FLOORPLAN image	6.6 (0.9)	20 (9.9)	17
4	ADD a new sound source (search, position and reach)	4.2 (1.6)	154 (64.5)	48
5	ADD a new sound source (search, position, reach and hidden)	6.6 (0.6)	62 (18.1)	38
6	ADD a new sound source (search, position, and fade)	6.6 (0.6)	82 (46.7)	43
7	Set all sound sources to START PLAYING WHEN ENTERING reach	5.6 (0.9)	70 (63.5)	25
8	Set sound sources TIMINGS to play in order	6.2 (0.5)	74 (60.1)	40
9	PLAY the soundscape and MOVE the listener around to listen to the soundscape	7.0 (0.0)	12 (11.0)	5
10	Add TAGS to the exhibition	5.8 (1.8)	50 (36.7)	15
11	SAVE and PUBLISH the exhibition	7.0 (0.0)	18 (2.9)	16
12	EXPORT the metadata to the laptop	6.8 (0.5)	14 (10.3)	8

We also observed difficulties when users were asked to set up the sound sources so that it would start playing when the user enters the area of reach. One participant struggled to find the drop-down menu and it was not clear to them what "start when entering" meant. One participant had difficulty seeing when a new sound source panel started and commented it was too easy to accidentally click on the delete button.

Finally, issues were observed when adding tags to the exhibition. One participant could not find the *Exhibition* tab, whereas another struggled to find the *Tags* field and checked throughout all tabs. Once found, one participant did not press Enter to add the tag as the call to action was not clear.

5.2. Part 2—Soundscape Curation

5.2.1. Methodology

In the second part participants were asked to become curators, and use what they had learnt about the Soundscape Create app in part 1 to recreate, as faithfully as possible, a virtual soundscape from a real one. For this purpose, we set up a room (rectangular—$14 \times 16 \times 6$ m) to look like it was part of an exhibition about space exploration. Specifically, we prepared an installation about the Voyager space program. We collaborated with a professional musician who composed and recorded a 5 min piece of music to use as accompaniment to speech excerpts from the *Golden Record* [46]. The content was mixed into 6 tracks, each of which was assigned to 1 of 6 loudspeakers placed around the room. A MaxMSP patch was used to control two laptops (a master and a slave) each one driving 3 loudspeakers. The patch allowed to assign tracks to speakers, set the volume of each track, control the playback—and used the Open Sound Control (OSC) protocol for communication between master and slave laptop.

Participants were first asked to listen and observe the real soundscape; they were also given a floor-plan of the room which they could use to take notes about the setup. Participants were asked to pay attention to the following: (1) match the track to the loudspeaker (each loudspeaker was labelled with the track assigned to it), (2) the loudspeakers' position (e.g., position in the room and height) and (3) the sound sources' reach (i.e., from how far the sound of each loudspeaker could be heard). Participants were allowed to listen to the soundscape as many times as needed, and after having noted down all the information, they needed to recreate the soundscape using the material (images and sounds) already available on the PLUGGY social platform. After completing the task, participants were asked to fill in the System Usability Scale (SUS) questionnaire [47] (Table 2) to measure usability. We used a 7-point Likert scale from *strongly disagree* (1) to *strongly agree* (7). Furthermore, to score how likely users would recommend PlugSonic to others, we adopted the net promoter score (NPS) [48] which is defined as the percentage of "promoters" (users rating their likelihood as 9 or 10 on a scale from 0 to 10) minus the percentage of "detractors" (6 or below); users rating 7 or 8 are considered "passive". Therefore, the NPS ranges from −100 (all detractors) to +100 (all promoters). We recruited 5 participants for part 2 (the same ones who participated in part 1).

5.2.2. Results

Overall, participants agreed that the application is easy to use and were confident in using it, but there is space for improvement with regards to clarity of features, iconography and the integration of features.

Specifically, participants found the visual and interactive aspect of PlugSonic most interesting as it gives quick results and a pleasant user experience. They also liked the possibility to test the soundscape and listen to the binaural audio directly on the Create app. What participants did not like about the application is that it first requires some level of understanding of the available features and icons. Furthermore, some tasks where judged as being "too manual" and requiring "too many clicks" to implement. Additional features were desirable such as being able to save the settings of a single sound and automatically close the source panel or choose the colour for the sources' icons.

Participants were neutral in whether they would use the system frequently. This is consistent with their rating of a 6 (out of 10) to the question how likely they would use any of the curation tools and add stories to the social platform. Although they see the novelty in the interactivity, the main reason for these scores is the fact that they do not work in the sector specifically or are not used to creating content.

For the NPS score: among the participants we had 1 detractor (one participant rated a 6), 2 passives (2 participants rated an 8) and 2 promoters (2 participants rated a 9). This results in an overall NPS score of 20, which is good. The list of questions with the average rating is shown in Table 2.

Table 2. Average system usability score for part 2 of the evaluation.

	Question	SUS Avg (Std)
1	I think that I would like to use this system frequently.	4.4 (1.7)
2	I found the system unnecessarily complex.	2.8 (1.3)
3	I thought the system was easy to use.	6.2 (0.8)
4	I think that I would need the support of a technical person to be able to use this.	3.0 (2)
5	I found the various functions in this system were well integrated.	5.4 (0.9)
6	I found functionality of features and controls clear.	5.0 (1.2)
7	I thought there was too much inconsistency in this application.	1.6 (0.6)
8	I would imagine most people would learn to use the application very quickly.	5.2 (1.6)
9	I felt very confident using the system.	5.8 (0.8)
10	I needed to learn a lot of things before I could get going with this application.	1.6 (0.6)
11	I wanted to share or talk to people about my experience with the application	5.8 (0.8)

5.3. Part 3—Soundscape Experience

5.3.1. Methodology

In part 3, participants were asked to use the PlugSonic Experience Mobile app to explore a soundscape within a real environment. The soundscape and physical space were the same used for part 2. The room was set up with photos and a video projection about the Voyager exploration program (Figure 6). Participants were asked to imagine visiting a museum and as part of the visit they were given a mobile device (iPad) which they could use to explore a soundscape. They were invited to explore the soundscape freely for as long as they wished while moving around the room. No specific recommendations were provided apart from the indication to point the device's camera in the direction they were facing—as the camera and inertial sensors were used by the device to infer position and orientation, which in turn were used by the mobile app to render the audio in 3D. They were also advised to ignore the device's screen as no information was going to be displayed. After exploring the soundscape, participants were asked to fill in a questionnaire about their emotional response—selecting among 11 options (indifferent, interested, uninspired, bored, excited, captivated, engaged, disappointed, satisfied, neutral and frustrated) or by adding other ones. They were also asked to answer questions about social potential and learning while using the app—using a 7-point Likert scale (with 1 *strongly disagree* and 7 *strongly agree*). We recruited 7 participants for part 3: 5 males, 2 females in their mid 20s to mid 30s.

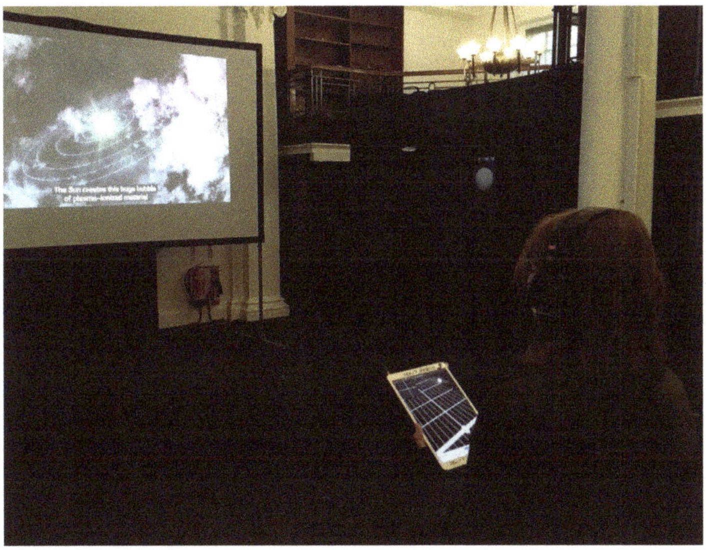

Figure 6. Participant using the Soundscape Mobile app for part 3 of the evaluation.

5.3.2. Results

During this experience, most participants have indicated feeling "interested" (7 participants), "engaged" (6 participants), and "captivated" (6 participants). Some indicated to have felt "excited" (2 participants), "indifferent", "satisfied", "inspired", "immersed", "calm", "overwhelmed" or "confused". The average score for personal resonance and emotional connection was 5.2 out of 7 (see Table 3a). The average score for learning and intellectual stimulation was 4.4 (see Table 3b). The average score for shared experience and social connectedness was 5.8 (see Table 3c). This means future improvements should be made to increase scores for personal resonance and emotional connection and learning and intellectual stimulation. It is important to underline how the soundscape adopted for the evaluation did not contain any speech or narration describing the material presented, which might help to explain the score for learning and intellectual stimulation.

Table 3. Evaluation part 3.

	a	
	Questions **Personal Resonance and Emotional Connection**	**Likert Score** **Avg (Std)**
1	The soundscape made experiencing the objects more interesting/fun.	5.4 (1.5)
2	I found the soundscape emotionally engaging.	5.6 (1.0)
3	During the experience, I felt connected with the objects presented to me	4.1 (2.0)
4	I will be thinking about this experience for some time to come.	5.7 (1.3)
	b	
	Learning and Intellectual Stimulation	
5	I got a good understanding about the objects presented to me.	3.1 (1.5)
6	I got a good understanding about where the objects where located.	4.9 (2.4)
7	I felt engaged with the objects presented to me.	5.0 (1.5)
8	I felt challenged and provoked.	4.0 (2.1)
9	During this experience, my eyes were opened to new ideas.	4.9 (2.0)
	c	
	Shared Experience and Social Connectedness	
10	I would have liked to have shared about it with other people	5.7 (1.1)
11	After this experience, I wanted to talk to people about it.	5.9 (0.9)

5.4. Discussion

The evaluation aimed at understanding if PlugSonic (1) is easy to learn and use for people unfamiliar with 3D audio in general and the concept of soundscape in particular, (2) is an effective tool for the creation of 3D soundscape from and for real environments and (3) has the potential to improve engagement and understanding of cultural heritage.

Overall, creating a soundscape was rated with a 6.4 (out of 7) for ease. Participants felt confident using PlugSonic Soundscape and found the interactive aspect of creating the soundscape most interesting and engaging. Some difficulties were observed, indicating that improvements could be made for adding and setting up new sound sources and tagging exhibitions. More in general, further improvements are possible to make the apps more streamlined and speed up the creation process. At the time of writing, the authors already updated PlugSonic Soundscape Create based on the feedback obtained during the evaluation. For example, the reach radius slider visibility has been improved; a visual feedback has been added when listener is in reach of a source (reach area changes colour); precision for all sliders has been set to one decimal point; the separation between sources' panels has been made clearer; and a "Loading.." message has been added when retrieving a sound source.

With a net promoter score of 20, participants are likely to recommend PlugSonic Soundscape to others. Although, one of the most critical aspects highlighted during evaluation was the resistance of users to become content creators. It seems clear that efforts are still necessary from research communities and institutions to empower citizens and provide them with more active participation in the act of sharing and protecting cultural heritage.

Additionally, for the experience part of PlugSonic Soundscape, positive results were obtained in terms of personal resonance as well as shared experience and social connectedness. Future improvements should be made to support emotional connection and intellectual stimulation better. Although, it needs to be taken into account that the evaluation did not take place in an actual museum with real artefacts being exhibited.

The main limitation in the evaluation of PlugSonic was in the number of participants. Only a total of 12 participants were involved, 5 for part 1 and 2 and 7 for part 3. Adding the baseline completion time in Table 1 helped to partially compensate for it. Although the apps would benefit from an extended testing and evaluation, similar tools [49] have been evaluated with similar number of participants (10).

We can conclude that, even accounting for its current limitations, the *PlugSonic Suite* represents a major contribution to the digital heritage community, including both end users and researchers. Our evaluation with inexperienced participants showed that the apps and the creation, curation and experience processes, within the support of the PLUGGY social platform, were easily understood and effectively utilised. The speed with which participants were able to familiarise and perform tasks with PlugSonic Soundscape Create shows how the design and implementation choices seem to have worked to simplify the learning process. Participants appreciated the aspect and the interaction with the app and did not feel limited during the recreation of a virtual soundscape from a real one. This was a particularly important aspect for the project, proving that 3D audio technologies can indeed be democratised without sacrificing the rendering quality and interaction flexibility. Furthermore, the Experience Mobile app was well received; there was no particular friction in the navigation of the soundscape within a real space. From the questionnaires' answers it seems clear how 3D audio soundscapes in general and PlugSonic in particular, can help cultural institutions in their mission to deliver engaging experiences and connect the public with cultural heritage.

6. Conclusions and Future Work

This paper presents the design, development and evaluation of a series of web and mobile applications—the *PlugSonic Suite*—for the curation and experience of 3D interactive soundscapes and audio narratives. These apps can be used to edit sound files and to create, test and experience soundscapes in a low friction environment that allows to transition from a web browser to a physical space navigation when desired.

The umbrella project (PLUGGY) includes the development of a social platform and several apps (AR, 3D Audio, Geolocation and Gamification) to provide users with the necessary tools to shape cultural heritage; both as curators and visitors of virtual or augmented exhibitions.

In Section 2, we defined sonic narratives and soundscapes, and introduced the binaural spatialisation technique and modern web-based technologies. We also described the state-of-the-art for sonic narratives, with special attention to cultural heritage applications. Even if research in these fields have produced significant advances, no specific tools have been developed to democratise spatial audio technologies and encourage the general public to adopt them to impact tangible and intangible heritage fruition and dissemination.

PlugSonic aims to demonstrate that spatial audio technologies and software can be designed to be accessible to anyone, without having to compromise on the binaural rendering quality or the flexibility of interactions available to the content creator, whether institutions or general public. After introducing the technology used for the development, we described the features and functionalities included in the apps—which were, partially, the result of a previous work with experts in the fields of cultural heritage and/or audio [45].

To understand whether our aims were achieved, we conducted an evaluation with subjects without previous experience and knowledge about 3D audio or soundscape design. The evaluation included three parts: creation, curation and experience of soundscapes with the PlugSonic Soundscape Create Web and PlugSonic Soundscape Experience Mobile applications. Participants were able to learn to use the apps quickly and effectively. They judged the task of creating a soundscape fairly simple and appreciated the level of interactivity and the possibility to test the results during the creation. Good results were obtained also for the task of curating a virtual soundscape starting from a real one, which shows how PlugSonic can indeed be effectively utilised to convey the atmosphere of a "real-world" situation. Limitations were highlighted when adding and setting up new sound sources and tagging exhibitions. A critical aspect observed in the evaluation was the resistance of users to become content creators. Further efforts seem necessary from research communities and institutions to empower citizens and allow them a more active participation in the act of sharing and protecting cultural heritage. The Experience Mobile app, exploiting the Apple ARkit [43] for the localisation of the user within a soundscape in a physical space, showed the potential of augmented exhibitions in increasing emotional resonance and connectedness with cultural heritage.

The feedback received during evaluation was used to improve user interface and user experience, but further work is necessary to simplify and speed up the soundscape creation process. The apps could also be improved by introducing moving sound sources, for an even more dynamic experience, and radiation patterns, for a more realistic simulation of sounds' directivity or to emulate occlusion effects. The Soundscape Create app already allows to choose the shape of the virtual environment, limited to rectangular and round spaces, and to use an image as floor-plan or background of a soundscape. Further improvements would include custom room shapes and the options to set movement constraints.

Furthermore, we think that PlugSonic has the potential to become also a web- and mobile-based 3D audio research evaluation tool. In fact, even if web-based audio evaluation tools are available [50,51], none focus specifically on spatial audio topics (e.g., HRTF selection, HRTF adaptation, speech reception threshold and cocktail party effect). Moreover, the hearing loss and hearing aid simulation algorithms available from the 3D TuneIn Toolkit would also allow for hearing impairment specific tests. PlugSonic could be used within a listening test framework that lets researchers easily design different types of online tests (e.g., AB, ABX and MUSHRA). Both the web- and mobile-based apps could be exploited for localisation tests or games within a virtual or physical space. To conclude, it is also worth highlighting how PlugSonic could continuously improve and be extended as a direct result of the constant development of the underlining technologies (WAA, 3D TuneIn Toolkit and ARKit).

Cultural institutions have started to adopt social media to promote events and stimulate participation, but cannot rely on a common ground when it comes to communication channels and tools to involve their audience. Being developed within the framework of the PLUGGY social platforms, PlugSonic could help to bridge the gap between general public and cultural institutions in an effort to encourage participation, co-creation and sharing of cultural heritage; especially because PlugSonic does not require the development or installation of software or hardware, and can be used on any device. Anyone could—visiting a museum or a monument—retrieve audio narratives from the social platform servers and experience them straight away.

7. Links

- PlugSonic Sample (standalone version): http://plugsonic.pluggy.eu/sample
- PlugSonic Soundscape Web (standalone version): http://plugsonic.pluggy.eu/soundscape
- Repository https://github.com/lpicinali/PlugSonic-soundscape
- PLUGGY social platform: https://pluggy.eu

(all links accessed on 8 February 2021)

Author Contributions: Conceptualization, L.P., M.C. and A.G.; methodology, M.C., L.P. and A.G.; software, M.C. and A.G.; validation, M.C. and L.P.; formal analysis, V.L. and M.C.; investigation, M.C.; resources, M.C. and L.P.; data curation, M.C. and L.P.; writing—original draft preparation, M.C.; writing—review and editing, M.C., A.G. and L.P.; visualization, M.C.; supervision, L.P.; project administration, L.P. and M.C.; funding acquisition, L.P. All authors have read and agreed to the published version of the manuscript.

Funding: This work was supported by the PLUGGY project (https://www.pluggy-project.eu/ (accessed on 8 February 2021)), European Union's Horizon 2020 research and innovation programme under grant agreement No 726765.

Institutional Review Board Statement: The study was conducted according to the guidelines of the Declaration of Helsinki, and approved by the Imperial College Research Ethics Committee (reference: 17IC4197; approved in October 2017).

Informed Consent Statement: Informed consent was obtained from all subjects involved in the study.

Data Availability Statement: Data supporting reported results: https://doi.org/10.5281/zenodo.4513876 (accessed on 8 February 2021).

Conflicts of Interest: The authors declare no conflict of interest.

Appendix A. PlugSonic Sample UI Controls Details

With reference to Figure A1, the user interface includes the following.

1. Waveform of the audio file with filename on the top left corner
2. Use of mouse to control the playback start and end points and select parts of the waveform to be modified
3. Play button to reproduce the whole audio file or the selected part
4. Stop button to stop reproduction
5. Undo button to cancel edit actions
6. Cut button to cut part of the waveform
7. Copy button to copy part of the waveform
8. Paste button to paste cut/copied part of the waveform
9. Mute button to mute part of the waveform
10. Fade In button to apply a volume fade in to the selection
11. Fade Out button to apply a volume fade out to the selection
12. Filters button to open the filters/effects menu (15)
13. Export button to save the modified audio file to the users device
14. Save button to save the modified audio file to the PLUGGY social platform
15. Audio Filters/Effects menu
16. Equaliser panel. Includes: lowpass, highpass, bandpass, lowshelf, highshelf, peaking and notch filters
17. Compressor effect panel with threshold, knee, ratio, attack and release controls
18. Reverb effect panel. Includes small, medium and large room reverbs

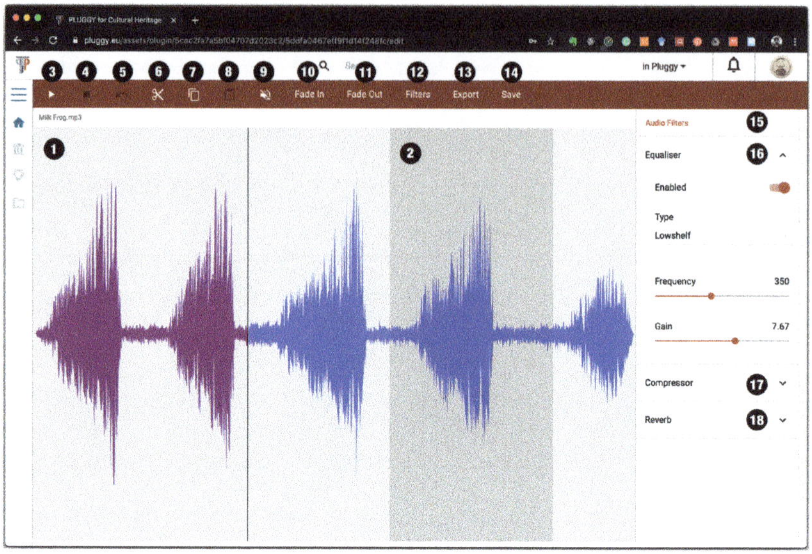

Figure A1. PlugSonic Sample user interface.

Appendix B. PlugSonic Soundscape Create UI Controls Details

With reference to Figure A2, the user interface includes:

1. Virtual room used for the curation of the soundscape
2. Listener's icon representing position and orientation of the listener in the virtual environment
3. Sound sources' icons representing position of each sound source in the virtual environment
4. Play button
5. Stop button
6. Record button to record and export a .wav file of the rendered 3D audio
7. Touch arrows button to open a panel showing touch arrow controls (necessary to navigate the soundscape on a touchscreen-based device)
8. Search tab. To search and retrieve audio sources from the social platform (Figure 3A):
    a. *Search* text field
    b. Dropdown menu to search among the user's audio assets (*My Assets*) or the whole social platform (*All Pluggy*)
    c. Dropdown menu to choose the ordering of the results
    d. *Search* button
    e. One button for each sound found by the search. Upon clicking a button the source is retrieved and added to the soundscape.
9. Sound sources' tab. Each sound source has the following options (Figure 3B):
    a. *On/off* toggle to activate/deactivate the sound
    b. *Volume* slider
    c. Position options:
        i. *Relative to listener* toggle. To set the sound's position in an absolute or relative fashion
        ii. Position sliders. *X/Y/Z* for absolute positioning or *Angle/Distance* for relative positioning
    d. *Loop* toggle to choose if the sound source will loop or play only once
    e. *Spatialisation* toggle to turn on/off the spatialisation engine. When off the source is reproduced as a mono/stereo file depending on the original file format

f. Reach options:

   i. *Reach* toggle—to turn on/off the interaction area. The interaction area (in yellow in Figure A2) is used to control the interaction between listener and sound source. When on, the listener will be able to hear the specific sound source only when they are inside the interaction area.
   ii. *Reach radius* slider—to choose the size of the interaction area
   iii. *Reach behaviour* dropdown menu—to choose the type of action the app will perform when entering/exiting the interaction area.
      - *Fade in and out*—playback will start as the user clicks the Play button but the source's volume will fade in/out as the listener enters/exits the interaction area
      - *Start when entering*—Playback will start as the listener enters the interaction area
   iv. *Fade duration* slider—to set the volume's fade in/out duration

g. *Timings* dropdown menu—to set an order in the reproduction of the sound sources. The reproduction of a specific source can be constrained to the start of another one.
h. *Hidden* toggle—to hide the sound source in the Soundscape Experience apps so that the user cannot see the source's position on the screen
i. *Delete* button—to delete the sound source from the soundscape

10. Room options tab. The room has the following options (Figure 3C):
   a. *Room Shape* dropdown menu (rectangular or round)
   b. *Room Size* text fields (Width/Depth/Height)
   c. *Room floorplan*—to search and select an image asset to be used as the soundscape's floor-plan.
   d. *Reset listener position* button—to reset the listener's position to coordinate (0, 0)
11. Listener's options tab—to set options regarding the 3D audio rendering engine. The listener's options are the following (Figure 3D):
   a. *Performance mode* toggle. When on, the performance mode is enabled, requiring less computational effort, allowing the rendering on low performance devices.
   b. *HRTF function* dropdown menu—to select the head related transfer functions to be used for the 3D sound rendering
   c. *HRTF sample length* dropdown menu—to choose between 128/256/512 samples long HRTFs
12. Exhibition tab—to set the exhibition's options, save and publish the exhibition (Figure A2). It includes:
   a. *Title* text field—to set the exhibition's title
   b. *Description* text field—to add a description of the soundscape
   c. *Tags* text field and icons—to add/delete tags to the exhibition
   d. *Save* button—to save the exhibition in the social platform
   e. *Publish/Unpublish* button—to make the exhibition available or not to the social platform's users.
   f. *Import* button—to import a previously exported soundscape in either format (metadata or metadata + assets)
   g. *Export* metadata button—to export the exhibition's metadata (will require access to PLUGGY social platform to retrieve the audio files)
   h. *Export* metadata+assets button—to export the exhibition's metadata and the audio assets in one file (will not require access to PLUGGY social platform to retrieve the audio files. Can be experienced when off-line)

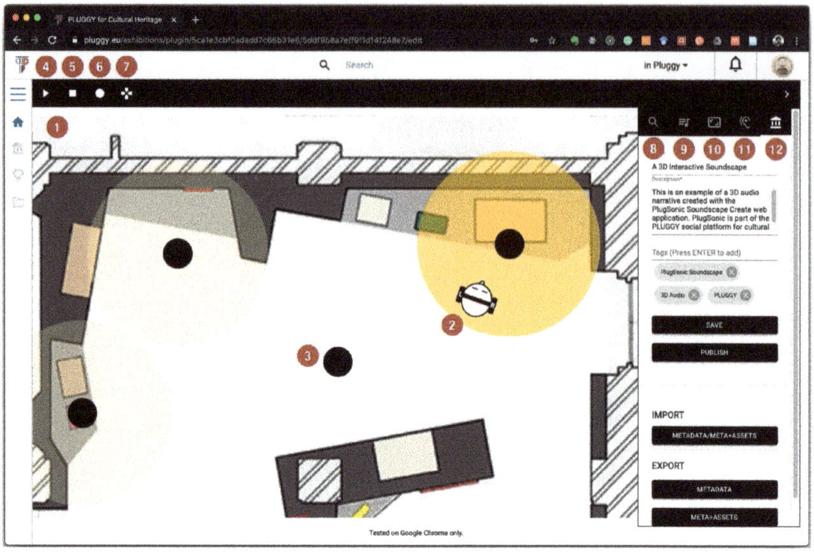

Figure A2. PlugSonic Soundscape Create user interface.

References

1. Fairclough, G.; Dragićević-Šešić, M.; Rogač-Mijatović, L.; Auclair, E.; Soini, K. The Faro convention, a new paradigm for socially - and culturally - sustainable heritage action? *Culture*; **2014**, *8*, 9–19. Available online: https://journals.cultcenter.net/index.php/culture/article/view/111 (accessed on 27 January 2021).
2. Russo, A. The rise of the media museum: Creating interactive cultural experiences through social media. In *Heritage and Social Media. Understanding Heritage in a Participatory Culture*; Routledge: London, UK, 2012; pp. 145–157. [CrossRef]
3. Stuedahl, D.; Mörtberg, C. Heritage knowledge, social media, and the sustainability of the intangible. In *Heritage and Social Media. Understanding Heritage in a Participatory Culture*; Routledge: London, UK, 2012; pp. 106–125. [CrossRef]
4. Lim, V.; Frangakis, N.; Tanco, L.M.; Picinali, L. PLUGGY: A pluggable social platform for cultural heritage awareness and participation. In *Advances in Digital Cultural Heritage*; Springer: Cham, Switzerland, 2018; pp. 117–129. Available online: https://doi.org/10.1007/978-3-319-75789-6_9 (accessed on 27 January 2021). [CrossRef]
5. Russo, A.; Watkins, J.; Kelly, L.; Chan, S. Participatory communication with social media. *Curator Mus. J.* **2008**, *51*, 21–31. Available online: https://onlinelibrary.wiley.com/doi/abs/10.1111/j.2151-6952.2008.tb00292.x (accessed on 27 January 2021). [CrossRef]
6. PLUGGY Project. Available online: https://www.pluggy-project.eu/ (accessed on 27 January 2021).
7. Europeana. Available online: https://www.europeana.eu/portal/en (accessed on 27 January 2021).
8. Katz, B.F.; Murphy, D.; Farina, A. The Past Has Ears (PHE): XR explorations of acoustic spaces as cultural heritage. In *Lecture Notes in Computer Science. Proceedings of the International Conference on Augmented Reality, Virtual Reality and Computer Graphics; Lecce, Italy, 7–10 September 2020*; Springer: Cham, Switzerland, 2020; pp. 91–98. Available online: https://doi.org/10.1007/978-3-030-58468-9_7 (accessed on 27 January 2021). [CrossRef]
9. Brezina, P. Acoustics of historic spaces as a form of intangible cultural heritage. *Antiquity* **2013**, *87*, 574–580. Available online: https://doi.org/10.1017/S0003598X00049139 (accessed on 27 January 2021). [CrossRef]
10. Dumyahn, S.L.; Pijanowski, B.C. Soundscape conservation. *Landsc. Ecol.* **2011**, *26*, 1327–1344. Available online: https://doi.org/10.1007/s10980-011-9635-x (accessed on 27 January 2021). [CrossRef]
11. Kytö, M.; Rémy, N.; Uimonen, H.; Acquier, F.; Bérubé, G.; Chelkoff, G.; Said, N.G.; Laroche, S.; McOisans, J.; Tixier, N.; et al. European Acoustic Heritage. 2012. Available online: https://hal.archives-ouvertes.fr/hal-00993848 (accessed on 27 January 2021).
12. Serafin, S.; Geronazzo, M.; Erkut, C.; Nilsson, N.C.; Nordahl, R. Sonic interactions in virtual reality: State of the art, current challenges, and future directions. *IEEE Comput. Graph. Appl.* **2018**, *38*, 31–43. [CrossRef] [PubMed]
13. Meelberg, V. Narrative sonic ambiances. Designing positive auditory environments using narrative strategies. In Proceedings of the Euronoise 2018, Crete, Greece, 27–31 May 2018; pp. 845–849. Available online: http://hdl.handle.net/2066/192407 (accessed on 27 January 2021).

14. Delle Monache, S.; Rocchesso, D.; Qi, J.; Buechley, L.; De Götzen, A.; Cestaro, D. Paper mechanisms for sonic interaction. In Proceedings of the Sixth International Conference on Tangible, Embedded and Embodied Interaction, Kingston, ON, Canada, 19–22 February 2012; pp. 61–68. [CrossRef]
15. Pijanowski, B.C.; Villanueva-Rivera, L.J.; Dumyahn, S.L.; Farina, A.; Krause, B.L.; Napoletano, B.M.; Gage, S.H.; Pieretti, N. Soundscape ecology: The science of sound in the landscape. *BioScience* **2011**, *61*, 203–216. Available online: https://doi.org/10.1525/bio.2011.61.3.6 (accessed on 27 January 2021). [CrossRef]
16. Krakowsky, T. Sonic storytelling: Designing musical spaces. *AdAge* **2009**. Available online: https://adage.com/article/on-design/sonic-storytelling-designing-musical-spaces/138028/ (accessed on 27 January 2021).
17. Collins, P. Theatrophone: The 19th-century iPod. *New Sci.* **2008**, *197*, 44–45. [CrossRef]
18. WebAudio API. Available online: https://www.w3.org/TR/webaudio/ (accessed on 27 January 2021).
19. WebGL. Available online: https://developer.mozilla.org/en-US/docs/Web/API/WebGL_API (accessed on 27 January 2021).
20. Web Audio API Specification Proposal. Available online: https://www.w3.org/2011/audio/drafts/1WD/WebAudio/ (accessed on 27 January 2021).
21. Poirier-Quinot, D.; Katz, B.F. The Anaglyph binaural audio engine. In Proceedings of the Audio Engineering Society Convention 144, Milan, Italy, 23–26 March 2018. Available online: http://www.aes.org/e-lib/browse.cfm?elib=19544 (accessed on 27 January 2021).
22. Carpentier, T.; Noisternig, M.; Warusfel, O. Twenty years of Ircam Spat: Looking back, looking forward. In Proceedings of the 41st International Computer Music Conference (ICMC), Denton, TX, USA, 25 September–1 October 2015; pp. 270–277. Available online: https://hal.archives-ouvertes.fr/hal-01247594 (accessed on 27 January 2021).
23. Musil, T.; Noisternig, M.; Höldrich, R. A library for realtime 3d binaural sound reproduction in pure data (pd). In Proceedings of the 8th international conference on digital audio effects (DAFX-05), Madrid, Spain, 20–22 September 2005; pp. 167–171. Available online: http://dafx.de/paper-archive/2005/P_167.pdf (accessed on 27 January 2021).
24. Cuevas-Rodríguez, M.; Picinali, L.; González-Toledo, D.; Garre, C.; de la Rubia-Cuestas, E.; Molina-Tanco, L.; Reyes-Lecuona, A. 3D Tune-In Toolkit: An open-source library for real-time binaural spatialisation. *PLoS ONE* **2019**, *14*, e0211899, doi:10.1371/journal.pone.0211899. Available online: https://doi.org/10.1371/journal.pone.0211899 (accessed on 27 January 2021). [CrossRef] [PubMed]
25. Ardissono, L.; Kuflik, T.; Petrelli, D. Personalization in cultural heritage: The road travelled and the one ahead. *User Model. User-Adapt. Interact.* **2012**, *22*, 73–99. Available online: https://doi.org/10.1007/s11257-011-9104-x (accessed on 27 January 2021). [CrossRef]
26. Not, E.; Zancanaro, M. Content adaptation for audio-based hypertexts in physical environments. *Hypertext'98: Second Workshop on Adaptive Hypertext and Hypermedia*; 1998; pp. 27–34. Available online: https://citeseerx.ist.psu.edu/viewdoc/summary?doi=10.1.1.57.1908 (accessed on 27 January 2021).
27. Petrelli, D.; Not, E. User-centred design of flexible hypermedia for a mobile guide: Reflections on the HyperAudio experience. *User Model. User-Adapt. Interact.* **2005**, *15*, 303–338. Available online: https://doi.org/10.1007/s11257-005-8816-1 (accessed on 27 January 2021). [CrossRef]
28. Benelli, G.; Bianchi, A.; Marti, P.; Not, E.; Sennati, D. HIPS: Hyper-interaction within physical space. In Proceedings of the IEEE International Conference on Multimedia Computing and Systems, Florence, Italy, 7–11 June 1999; pp. 1075–1078. Available online: https://doi.ieeecomputersociety.org/10.1109/MMCS.1999.778663 (accessed on 27 January 2021). [CrossRef]
29. Zimmermann, A.; Lorenz, A. LISTEN: A user-adaptive audio-augmented museum guide. *User Model. User-Adapt. Interact.* **2008**, *18*, 389–416. Available online: https://doi.org/10.1007/s11257-008-9049-x (accessed on 27 January 2021). [CrossRef]
30. Delerue, O.; Warusfel, O. Authoring of virtual sound scenes in the context of the Listen project. In Proceedings of the Audio Engineering Society Conference, 22nd International Conference: Virtual, Synthetic, and Entertainment Audio, Espoo, Finland, 15–17 June 2002.
31. Zimmermann, A.; Lorenz, A.; Birlinghoven, S. Listen: Contextualized presentation for audio-augmented environments. In Proceedings of the 11th Workshop on Adaptivity and User Modeling in Interactive Systems, Bonn, Germany, 6–8 October 2003; pp. 351–357.
32. IRCAM Spatialisateur. Available online: http://forumnet.ircam.fr/product/spat-en/ (accessed on 27 January 2021).
33. Vayanou, M.; Katifori, A.; Karvounis, M.; Kourtis, V.; Kyriakidi, M.; Roussou, M.; Tsangaris, M.; Ioannidis, Y.; Balet, O.; Prados, T.; et al. Authoring personalized interactive museum stories. In *Lecture Notes in Computer Science, Proceedings of the International Conference on Interactive Digital Storytelling; Singapore, 3–6 November 2014*; Springer: Cham, Switzerland, 2014; pp. 37–48. Available online: https://doi.org/10.1007/978-3-319-12337-0_4 (accessed on 27 January 2021). [CrossRef]
34. Pujol, L.; Katifori, A.; Vayanou, M.; Roussou, M.; Karvounis, M.; Kyriakidi, M.; Eleftheratou, S.; Ioannidis, Y. From personalization to adaptivity: Creating immersive visits through interactive digital storytelling at the acropolis museum. In Proceedings of the 9th International Conference on Intelligent Environments, Athens, Greece, 16–19 July 2013; pp. 541–554. Available online: https://doi.org/10.3233/978-1-61499-286-8-541 (accessed on 27 January 2021). [CrossRef]
35. Hansen, F.A.; Kortbek, K.J.; Grønbæk, K. Mobile urban drama: Interactive storytelling in real world environments. *New Rev. Hypermedia Multimed.* **2012**, *18*, 63–89. Available online: https://doi.org/10.1080/13614568.2012.617842 (accessed on 27 January 2021). [CrossRef]
36. Emotive. Available online: https://emotiveproject.eu/ (accessed on 27 January 2021).

37. Roussou, M.; Ripanti, F.; Servi, K. Engaging visitors of archaeological sites through "emotive" storytelling experiences: A pilot at the ancient agora of Athens. *Archeol. E Calc.* **2017**, *28*, 405–420. Available online: https://doi.org/10.19282/AC.28.2.2017.33 (accessed on 27 January 2021). [CrossRef]
38. Arches. Available online: https://www.arches-project.eu/ (accessed on 27 January 2021).
39. Geier, M.; Spors, S. Spatial audio with the soundscape renderer. In Proceedings of the 27th Tonmeistertagung—VDT International Convention, Cologne, Germany, 22–25 November 2012. Available online: https://www.int.uni-rostock.de/fileadmin/user_upload/publications/spors/2012/Geier_TMT2012_SSR.pdf (accessed on 27 January 2021).
40. Vaananen, R. User interaction and authoring of 3D sound scenes in the Carrouso EU project. In Proceedings of the Audio Engineering Society Convention 114, Amsterdam, The Netherlands, 22–25 March 2003.
41. Echoes. Available online: https://echoes.xyz/ (accessed on 27 January 2021).
42. Comunità, M.; Gerino, A.; Lim, V.; Picinali, L. PlugSonic: A web- and mobile-based platform for binaural audio and sonic narratives. *arXiv* **2020**. Available online: https://arxiv.org/abs/2008.04638 (accessed on 27 January 2021).
43. Apple AR Kit. Available online: https://developer.apple.com/augmented-reality/ (accessed on 27 January 2021).
44. Faro Convention. Available online: https://www.coe.int/en/web/conventions/full-list/-/conventions/treaty/199 (accessed on 27 January 2021).
45. Comunità, M.; Gerino, A.; Lim, V.; Picinali, L. Web-based binaural audio and sonic narratives for cultural heritage. In Proceedings of the Audio Engineering Society Conference: 2019 AES International Conference on Immersive and Interactive Audio, York, UK, 27–29 March 2019, Available online: http://www.aes.org/e-lib/browse.cfm?elib=20435 (accessed on 27 January 2021).
46. Voyager Golden Record. Available online: https://voyager.jpl.nasa.gov/golden-record (accessed on 27 January 2021).
47. Lewis, J.R. The system usability scale: Past, present, and future. *Int. J. Hum. Comput. Interact.* **2018**, *34*, 577–590. Available online: https://doi.org/10.1080/10447318.2018.1455307 (accessed on 27 January 2021). [CrossRef]
48. Reichheld, F.F.; Covey, S.R. *The Ultimate Question: Driving Good Profits and True Growth;* Harvard Business School Press: Boston, MA, USA, 2006; Volume 211.
49. Çamcı, A.; Lee, K.; Roberts, C.J.; Forbes, A.G. INVISO: A cross-platform user interface for creating virtual sonic environments. In Proceedings of the 30th Annual ACM Symposium on User Interface Software and Technology, Québec City, QC, Canada, 22–25 October 2017; pp. 507–518. Available online: https://doi.org/10.1145/3126594.3126644 (accessed on 27 January 2021). [CrossRef]
50. Jillings, N.; Man, B.; Moffat, D.; Reiss, J.D. Web audio evaluation tool: A browser-based listening test environment. In Proceedings of the 12th International Conference on Sound and Music Computing, Maynooth, Ireland, 26 July–1 August 2015; pp. 147–152. Available online: https://doi.org/10.5281/zenodo.851157 (accessed on 27 January 2021). [CrossRef]
51. Kraft, S.; Zölzer, U. BeaqleJS: HTML5 and JavaScript based framework for the subjective evaluation of audio quality. In Proceedings of the Linux Audio Conference, Karlsruhe, Germany, 1–4 May 2014. Available online: http://lac.linuxaudio.org/2014/papers/26.pdf (accessed on 27 January 2021).

applied sciences

MDPI

Article

Towards a More Accessible Cultural Heritage: Challenges and Opportunities in Contextualisation Using 3D Sound Narratives

Veranika Lim [1,*,†], Sara Khan [2,†] and Lorenzo Picinali [1,*]

1 Dyson School of Design Engineering, Imperial College London, London SW7 2DB, UK
2 Department of Architecture and Design, Politecnico di Torino, 10129 Torino, Italy; khan.saretta@gmail.com
* Correspondence: v.lim@imperial.ac.uk (V.L.); l.picinali@imperial.ac.uk (L.P.)
† These authors contributed equally to this work.

Abstract: This paper reports on the exploration of potential design opportunities for social media and technology to identify issues and challenges in involving people in generating content within a cultural heritage context. The work is divided into two parts. In the first part, arguments are informed by findings from 22 in-depth semi-structured interviews with representatives of cultural institutions and with people from a general audience who recently participated in a cultural activity. The key findings show that social media could be used more extensively to achieve a deeper understanding of cultural diversity, with opportunities in *redefining the expert*, *extending the experience space*, and *decentralising collaboration*. To further support these findings, a case study was set up evaluating the experience of a mini audio tour with user-generated (i.e., personal stories from a local audience) vs. non user-generated (i.e., professional stories including facts) narratives. These were delivered using text and 3D sound on a mobile device. The narratives were related to a built environment in central London near world-renown museums, cultural buildings, and a royal park. Observations, a standardised spatial presence questionnaire, and a short open interview at the end of the tour were used to gain insights about participants preferences and overall experience. Thematic analysis and triangulation were used as a means for understanding and articulating opportunities for social media to better involve and engage people using user-generated narratives presented through 3D sound.

Keywords: cultural heritage; sound and narrative design; built environment; user-generated content

Citation: Lim, V.; Khan, S.; Picinali, L. Towards a More Accessible Cultural Heritage: Challenges and Opportunities in Contextualisation Using 3D Sound Narratives. *Appl. Sci.* **2021**, *11*, 3336. https://doi.org/10.3390/app11083336

Academic Editor: Marek Milosz
Received: 6 March 2021
Accepted: 31 March 2021
Published: 8 April 2021

1. Introduction

Living in the era of an always-connected society has created great opportunities for developing technological tools to support distributed curation in cultural heritage. The most known "place" for distributed curation is social networks, also called user-generated content (UGC). In the last decade, an increasing number of cultural heritage venues have started to adopt the use of social networks to facilitate cultural contributions from everyday visitors [1]. As a result, there is an continuously increasing amount of UGC on the Web.

Despite the "explosion" of UGC, only a small portion of people contribute to generating new content. There is an emerging rule of thumb that suggests that, if you get a group of 100 people online, then one will create content, 10 will "interact" with it (commenting or offering improvements), and the other 89 will just view it [2]. Furthermore, although social media is currently being used by many cultural institutions to reach their audiences, our findings show that there is one mayor issue; the large amount of information provided and available online is problematic for an efficient reach to potential audiences. To our knowledge, however, distributed curation has not been fully exploited with regard to cultural heritage promotion and integration in people's everyday lives.

This paper reports on an exploration of potential design opportunities for social media and technology to identify issues and challenges in involving people in distributed curation through 2 studies. In the first study, we took an empirical and critical approach using data from in-depth, semi-structured, and audio-recorded interviews to identify potential design opportunities. In the second study, we aimed to support the interview results using a case study. In the case study, we explored how user-generated content could contribute to visitor's engagement around built heritage in an "everyday life" context using 3D sound narratives. Overall, this work is conducted as part of a larger project called *PLUGGY* [3], which aims to allow users to share their local knowledge and everyday experience with others, and alongside cultural institutions. PLUGGY aims to build extensive networks around a common interest around cultural heritage.

In the following sections, we first introduce an emerging paradigm shift to contextualise our approach. We show a number of recent examples of projects in Human–Computer Interaction (HCI) that aimed to support distributed curation using social media. We then present the methods and outcomes of a series of in-depth semi-structured interviews. Themes are discussed on how existing social media could be better utilised to improve user engagement and to encourage distributed curation within a cultural heritage context. We then present the case study where we evaluated the experience of a mini audio tour with user-generated (i.e., personal stories from a local audience) vs. non user-generated (i.e., professional stories including facts) narratives.

2. Background

2.1. A New Paradigm in Cultural Heritage

In the traditional view of cultural heritage, people valued material things with intrinsic properties or a national history relevance. The traditional paradigm encouraged the reduction of heritage to tourism and consumption, with the engagement of broader cultural topics only for a professional audience. Recently, however, a new paradigm emerged that puts the production of heritage to the foreground. The new paradigm aims to encompass the greater involvement of local citizens in creating content and to contribute to cultural heritage, with greater concern for the everyday. Generally, it has been argued that, when heritage is everywhere and relevant to our everyday life, it is likely to be one of the preconditions for genuine sustainability [4]. This stresses the importance of heritage communities, which are social groups, who value specific aspects of cultural heritage that they wish to sustain and transmit to future generations within the framework of public action [5] and social capital, which is defined as an important resource of individuals and social groups impacting economic growth, democratic practices, quality of governance, and quality of life [6]. Hence, local citizen participation is considered an ethical obligation and a political necessity to strengthen democracy and to create governance that can ultimately improve quality of life [4].

Some key examples of enablers of the paradigm shift are *policy*, *inclusivity*, and *technology*. Firstly, the role of culture as a component of sustainable development is being increasingly discussed in policy debates. UNESCO emphasised the importance of culture during the Decade of Culture and Development (1988–1998) and through its conventions (e.g., on the Protection and Promotion of the Diversity of Cultural Expressions in 2005, for the Safeguarding of the Intangible Cultural Heritage in 2003, and concerning the Protection of the World Cultural and Natural Heritage in 1972). In Europe, both the Council of Europe's Landscape Convention and its Faro Convention imply the need for more culturally sensitive approaches [4].

Secondly, heritage is often represented by places linked to a built environment that is already iconic but not really representative of all the sides and transliterations of a culturally diverse audience [7]. This happens also on a smaller scale in museum environments, e.g., the Mona Lisa in the Louvre museum is one of the most famous masterpieces, but most of the time, people focus just on this artwork and easily forget to pay attention to the surroundings. According to a study published by the MIT Senseable Lab [8], generally, visitors overlook the very famous Giotto's display just in front of da Vinci's masterpiece. The reason for this is because anyone who visits the museum feels the need to snap a picture of the Mona Lisa, as a means to underline the *"Louvre museum experience"* and *"to bring back home the proof of having been there"* as a small souvenir. Therefore, collecting samples of the museum/cultural experience is a recurring action for tourists and museum visitors [9], while cultural landmarks are becoming similar to brands. To reach better visibility, cultural heritage needs to be understood by a wider range of visitors. One key aspect in being understood by the audience is to find a common root between different cultures and to globalise what, at the moment, is only locally known [9]. The challenge here is to contextualise masterpieces, cultural landmarks, and pieces of the built environment to a culturally diverse panorama. Cultural artefacts are rarely located where most people live or work: *here*. Too often they are—it seems almost by definition—somewhere else: *there*. People might visit them on holiday, but this type of heritage is not part of the everyday landscape of their normal lives. If heritage is thus defined as elsewhere, there is a risk that it will unintentionally become an instrument of exclusion. It can only become an instrument of inclusion and commonality if it is defined contextually as local, lived-in, and ordinary and if it is recognised as an element of both shared identity and differentiation [10]. According to Giaccardi, contemporary heritage studies are the results of ongoing interactions in the lived world of ordinary people [11].

Thirdly, because of careful planning of mobile technology manufacturers and telecommunications service providers, we have now reached an always-connected society [12]. The combination of ubiquitous access to multimedia content and information with the consolidation of distributed curation generates great opportunities to develop technological tools to further enable this paradigm shift in cultural heritage.

2.2. Distributed Curation in Social Media and HCI

Distributed curation is a phenomenon by which citizen communities have accepted freely providing content and information [13]. Digital and social technologies are facilitating distributed forms of curatorial practice, which can be harnessed to democratise history [14], in turn supporting inclusivity and policy. Liu [14] describes distributed curation as a sociotechnical practice involving people, cultural artefacts, and information and communication technology. It is a collaborative and distributed practice, creating shared ownership over the stewardship of the living heritage through transparency, which further allows other parties to partake in the curatorial process.

There have been a number of recent projects in HCI that aimed to support distributed curation using social media. One well-known social media channel generally accepted and used by communities is Instagram. Instagram allows creative practices from non-elite social contexts and communication that relies on everyday competencies rather than on formal artistic education (Burgess, 2006 as described in [15]), extending the reach of the museum beyond its walls. An example is the *CURIOS* project [1]. *CURIOS* researchers explored how digital archives for rural community heritage groups can be made more sustainable so that volunteer members can maintain a lasting digital presence. They developed software tools to help remote rural communities collaboratively maintain and present information about their cultural heritage using open linked data. This approach is adjusted to the specificity of a local heritage but can also take advantage of already collected materials from elsewhere. In another example, Yelmi et al. [16] evaluated soundscapes as an intangible cultural heritage element and introduced the *Soundsslike* project. *Soundsslike* is a crowd sourced online sound archive that invites people to record symbolic urban sounds and to upload them to an online sound archive. This online platform was built and displayed in an exhibition through an interactive table-top interface to learn more from users and contributors and to enrich the archived content by raising public awareness of urban sounds.

Distributed curation has also been shown to be helpful in personalising experiences. For example, *MobiTag* [17] is an electronic guide that supports semantic, social, and spatial navigation in museums by allowing visitors to create and vote for tags. Han et al. [18] developed a mobile application called *Lost State College* (LSC) and showed that users used social features as a way of learning local history and of interacting with others, co-creating digital traces and rich layers of local history information. McGookin and Brewster designed

PULSE [19], which allows users to gain a vibe (i.e., an intrinsic understanding of the people, places, and activities around their current location) using Twitter data. As users moved, *PULSE* downloaded public messages (tweets) generated by any user in the current location. Then, *PULSE* selected the closest tweet and inserted it in a virtual 3D auditory environment: users heard tweets as whispered conversations. Bellens et al. [20] explored how social media data can be employed to study tourism on European cultural routes and showed its potential for investigating a complex touristic object such as a cultural route. They combined text related to photos on Instagram with Wikipedia for geographical places. This allowed them to identify the most popular stops and localities related to the cultural route. In addition, Bujari et al. [21] proposed *PhotoTrip*, an interactive tool able to autonomously recommend cultural heritage locations along travel itineraries even if those locations were not mainstream. *PhotoTrip* identified points of interest by gathering pictures and related information from Flickr and Wikipedia and then by providing the user with suggestions and recommendations.

Through these examples, we can see that communities are being more and more involved through distributed curation, where technologies mediate and allow users to contribute to our histories. This supports what was found in previous work, claiming that awareness towards local heritage should be promoted at first among locals in their living or working environment [22–24]. The results of these projects and of the current use of social platforms have demonstrated their potential to build networks through the individual and distributed contributions of users. To our knowledge, however, these possibilities have not been fully exploited with regard to cultural heritage promotion and integration in people's everyday life. We believe that distributed curation is still limited in the cultural heritage sector. Most of the time, tourists deal with specific tourism-related agents and locations, such as tour guides, desk clerks, and taxi drivers, and rarely merge with the surrounding community [25]. Mass tourism can create a barrier between visitors and locals. It is therefore not surprising that a wide range of new concepts such as "sustainable tourism", "resident responsive tourism", and "community-based tourism" have become important. Visitors are in constant need of help in finding relevant information, but providing them with appropriate information is challenging because their interests and needs are unknown [26].

This paper reports on two studies we conducted to explore potential design opportunities for social media and technology and to identify issues and challenges in empowering people to participate in cultural heritage curation. The first study took an empirical and critical approach using data from in-depth, semi-structured, and audio-recorded interviews. In the second study, we aimed to support the results coming from the interviews through a case study. For the case study, we looked at broadening the narrative scope of a built environment using curator vs. user-generated 3D audio narratives. The aim was to explore how user-generated content could contribute to visitor's engagement within an "everyday life" context, and to better understand the challenges we face and the opportunities we can utilise to enhance engagement.

3. Study 1: In-Depth Interviews

In-depth interviews are the cornerstone of design thinking. Through research techniques such as in-depth interviews, we can learn how different users or customers feel about the problem one is trying to solve and how they might fix it if they could.

3.1. Methodology

3.1.1. Participants

We conducted in-depth semi-structured interviews with 22 participants who could potentially be early adopters of PLUGGY. The participants were divided into two groups: cultural heritage professionals and a general audience who recently participated in cultural activities. Heritage professionals were recruited through our immediate social networks and referrals working in the cultural heritage industry. For the general audience target, we recruited connections within our social network who have recently visited a cultural heritage site. Interviewing two different stakeholder groups allowed us to see the different perspectives in motivations, challenges, and design opportunities. Cultural heritage professionals varied in occupation including museums' head of media, curators, guides, administrators, and managers; an architect; an economist; and a journalist. All worked in either museums, galleries, or independently, but all within the field of cultural heritage. The general audience participants included post-graduate students, researchers, a developer, a company founder, an account assistant, and a manager, who all often engage in and have recently engaged in leisure cultural heritage activities.

3.1.2. Procedures

Cultural heritage professionals were asked questions covering themes such as their background, recent projects and their motivations and goals, their social media and social technology use, and the main challenges in these. Other key themes were related to their views of what the future should bring in terms of experiences and the role of social technology and social media. General audience participants were asked questions covering themes such as personal background, their recent cultural heritage-related activities, how they use social media in their activities, their motivation and goals, and what challenges they commonly face. See Appendix A for the complete list of questions. The interviews lasted 30 to 90 min in Q3 of 2018 and took place in participants' respective locations or online using Skype. Interviews were audio recorded for transcription purposes.

3.1.3. Analysis

We conducted a thematic analysis in which we analysed the interviews in steps, gradually identifying emerging themes in the data. After the interviews were transcribed verbatim, items (i.e., paragraphs) were selected when relevant to our goals (i.e., highlighting opportunities to engage people better in the cultural heritage sector and exploring how current social media could further be utilised). Relevant items were then coded individually and discussed collaboratively by two authors. The MAXQDA (https://www.maxqda.com/, accessed on 26 March 2021) qualitative analysis software was used to map the frequency and relevance of codes. Based on these, we identified 3 main overarching themes, which are discussed in the next section using pseudonyms to refer to participants.

3.2. *Results*

We collected and transcribed a total of 930 min of audio-recorded interviews (approximately 27,500 words). A total of 283 comments were identified (one comment can be a sentence or a paragraph describing, for example, an issue, reflection, or concern); 165 comments were categorised to fit within 3 overarching themes selected for discussion in this paper. Comments that were mentioned only once were not considered as empirically relevant, and they were therefore left out of the analysis.

Overall, we identified three overarching themes, each identifying opportunities or implications for further exploration. These three themes are *redefining the expert*, *extending the experience space*, and *decentralising collaboration*. For each theme, we discuss the participants' main challenges, with an attempt to identify potential design opportunities for social media to promote cultural heritage and to foster cultural diversity.

3.2.1. Theme 1: Redefining the Expert

Firstly, our findings suggest considering local audiences as experts and, therefore, as an important resource and potential marketing channel. This adds to Kidd's argument about the relevance of promoting the "face" of an institution through local visitors or local residences, who should play a more significant role in creating an institution's face in promoting cultural heritage [27].

The following findings are also in support of this argument. Both cultural heritage professionals and general audience interviewees shared challenges related to finding content and events (related to 23 and 39 comments, respectively). Participants felt biased in discovering culture-related content as well as no sense of fulfillment in finding content at all. They mentioned being aware of the results of the algorithms to dictate what we encounter online. This, also called *filter bubble*, represents a state of intellectual isolation that can result from personalised searches when a website algorithm selectively guesses what information a user would like to see based on information about the user, such as location, past click behaviour, and search history [28]. General audience interviewees, for example, expressed the feeling of being overwhelmed or overloaded by the events, activities, and content promoted on social media pages such as Facebook. Specifically, events and content related to dance culture and folklore (i.e., intangible heritage) are mentioned as an even bigger challenge for discovery. This is understandable as Bea, a cultural heritage professional we interviewed, explains that *"almost every cultural institute, if not all, are reaching their audiences using similar standard means such as Facebook, Twitter, and Instagram"*. Likewise, Fabio, who is a journalist in Greece, seemed to have difficulties in reaching out to potential audiences in other countries such as Germany because he believes that the more audiences he has in Greece, the more his content is automatically shown to potential audiences in Greece.

In response to these challenges, general audience interviewees rely strongly on information from their immediate friends or from locals nearby during their cultural visits, as they have low confidence and trust in social media content for the reasons described above (related to 18 comments). These kinds of mouth-to-mouth discoveries usually take place by direct messaging, face to face, or when friends take one another on a cultural visit. Interviewees expressed their preference in approaching locals or connections as they hold immediate and trustful expertise of information beyond what is publicly promoted.

Examples of related work focusing on local contributions to cultural heritage can be found in the application developed by Han and colleagues [18] called *Lost State College* (LSC). This research shows that users employed social features as a way of learning local history and interacting with others, co-creating digital traces and rich layers of local history information. Another example is the *CURIOS* project [1], where researchers explored how digital archives for rural community heritage groups can be made more sustainable so that volunteer members can maintain a lasting digital presence. Something similar but for sound was introduced by Yelmi and colleagues [16]. Another platform named *ArtLinks* was developed to provide a guidance system based on a public display in museum exhibits, which allowed visitors to create and use geo-tags to help guide other visitors [21]. Similarly, *MobiTag* [17] is an electronic guide that supports semantic, social, and spatial navigation in museums by allowing visitors to create and vote for semantic tags to describe objects.

Although these examples are headed towards the right direction in redefining an archetypal expert, most of these studies were done within the conventional museum space. We believe that current social media and technologies do not sufficiently support local audiences in taking the role of the "expert", utilising their knowledge significantly in the everyday space. In order to do this, we need to extend the experience space beyond museums, archives, or specific heritage sites and allow local audiences to contribute and share heritage-related content easily.

3.2.2. Theme 2: Extending the Experience Space

Secondly, our findings show an opportunity in blending discovery and experience, e.g., what happens at home or in a hotel room could transition to what happens on site. Dialogue in social media often comes from a small number of contributors or active community members [29]. To encourage and enable more genuine participation, dialogue, and inclusive spaces, content could be made more dynamic and interesting for a broader audience who are navigating in a broader space. Kidd [27] discussed two examples of good practice for inclusive spaces, highlighting the need to accommodate different audiences. Examples of institutions that went in this direction are the Powerhouse Museum in Sydney, which has relaxed its policy, recognising that taking photographs is an important part of the experience for visitors. On a similar pathway, the Australian Museum started making a distinction between the needs of different kinds of "visitors" by hosting separate blogs. When aiming at further promoting inclusive spaces beyond museum spaces, our findings show an opportunity in blending discovery and experience.

Interviewees usually plan high-level activities and allow space for discoveries on site and in the moment. They tend to use cues within their environment (e.g., large colourful objects or crowds) to lead their attention and to make impromptu decisions. This is a great thing according to our professional interviewees, as it allows for serendipity. However, serendipity may decrease visitors' motivation to download an application that can be used in an exhibition or museum to enforce engagement. There are challenges in asking visitors to use their devices. They may face barriers such as limited internet access, limited time in understanding the app's features, or not enough memory space to download yet another app they may not use again after their visit. According to our interviewee Alfred, who is responsible for media, visitors do not want to download something for a visit of a couple of hours and then never use it again. Consequently, visitors may just come and see beautiful objects with limited engagement. Because of this, museums or cultural spaces are not perceived as part of visitors' everyday space. Technology limitations may making it difficult to transition between spaces because applications that are used on an everyday basis are not connected to museum or cultural spaces whatsoever.

Cultural heritage professional interviewees have focused too much on experiences inside a cultural institution or site as a way to improve visitor engagements with objects. However, they acknowledge the need to better connect objects with audiences' personal everyday lives off site (related to 34 comments). They emphasised the importance of contextualising artefacts to make them more relevant to audiences as part of their everyday reality and ideological viewpoint (e.g., how visitors can personally relate to an artefact displayed in a museum). This is considered crucial to generate opinions and engagements [26].

Making objects more relevant to audiences as part of their everyday reality means that cultural institutions need to reach their audiences outside their spaces. For example, audiences could be reached ahead of a visit as part of an (onsite) experience. Early discovery may also create a better understanding of cultural content once audiences are inside a dedicated cultural heritage space. Streamlining the discovery and experience of cultural heritage could be an avenue for exploration, leading towards an online multimedia museum with connections to the physical world. *PLUGGY* (Pluggable Social Platform for Heritage Awareness and Participation) is a recent project [3] aimed at doing this. *PLUGGY* is a social platform with curatorial tools to allow citizens to create personalised stories and exhibitions online and to share them through social networks with friends, associates, and professionals. In creating these online exhibitions accessible anywhere, users of *PLUGGY* can add their own assets as well as assets from public libraries and museum libraries. Another project that has explored the use of current leading social media in discovering culture beyond museum walls is *PULSE* (2012), an auditory display of twitter data created by McGookin and Brewster to allow users to gain a vibe or understanding of the people, places, and activities that occur in the user's current locale.

Another opportunity we foresee in extending the experience space is by facilitating documentation after events, by sharing this with the public, and by making it easier for visitors to provide feedback or to continue discussions (i.e., related to 11 comments). Findings show that activities are usually documented and shared in close networks of museum friends and, to some extent, using common social media channels such as Facebook and Twitter. Smaller venues such as galleries may not provide follow ups to events at all because of limited resources; factual or visual information about how was an event are either for personal records or for internal purposes and rarely shared with the public. Therefore, the emergence of digital means such as social media allows visitors to start discussions related to their visits in online public spaces, but these discussion may not always reach cultural venues. As a consequence, relevant key players who are responsible for delivering the experience being discussed online remain unaware about these discussions.

In some occasions, however, cultural experiences should not be shared on social medial at all. For example, for intangible cultural heritage such as traditional dances, the actual experience cannot fully be conveyed online and social media can only be a means to share and find events. According to Oscar *"Participation is as close as a 'truthful' experience, where one should travel to where it was created"*. Quin adds that *"People hold different interpretations of a dance which has lead to traditional dances going through a constant transformation from generation to generation"*. Hence, the value of intangible heritage lies in immediate and offline interactions. This should be taken into account when creating digital experiences.

3.2.3. Theme 3: Decentralising Collaborations

Thirdly, based on our findings, collaboration should be decentralised and content should be reflected upon by a general audience. Cultural institutions are often not able to provide reflections on user-curated contributions. Once contributions are hosted in institutions online, there is no guidance in making them "useful". According to Alfred, *"when visitors come with objects to his museum there is usually no strategy for evaluating them"*. Hence, we have not reached the goal of "becoming social" that has been set within the new *museology* introduced by Kidd [27].

Allowing a general audience to reflect on cultural content, however, requires a robust and reliable systematic review of who can be considered an expert. A main advantage with this approach is that it would allow for different perceptions on cultural content (linked to 30 transcription items). Our museum professionals acknowledge that they are not necessarily experts in all their collections. To solve this, they involve larger numbers of people to help understanding and curating collections. However, there remains a validation challenge. Moreover, validating the accuracy of audiences' contributions is even a bigger challenge. Different cultures may have conflicting stories, with none being objectively right or wrong. Additionally, there is no standard way or infrastructure to validate them.

Previous research showed that content created by a museum might not be as engaging as content created by others [30]. Involving visitors, locals, or the general audience in interpreting collections and in the validation process of content could in fact lead to a more social structure and, hence, a better understanding and appreciation of objects. Currently, most objects in a museum do not really have that voice. The language in museums prevents a full understanding of the artefact and hence opinions being created.

The collective creation of experiences by users is usually referred to as co-experience [31]. Empowering collaboration in generating content creates opportunities to connect objects and people. Grounding objects better within their original communities and environments is expected to allow visitors to gain "a greater worldview or a new set of connections and correlations between various dimensions of experiences" [32]). This is expected to lead to transformational experiences, which happen if we discard old ways of thinking and provide new opportunities for individuals to invent knowledge, transform what they have encountered in the past, and contribute to new ideas and concepts.

According to Batterbee and Koskinen [33], people design and create experiences collectively on a daily basis. However, co-experience is not about creating products or art; it is about the ways in which participants make things meaningful for each other. One example of co-experience is the *Curarium*, a digital interactive platform created by metaLAB at Harvard. *Curarium* employs crowd sourcing to annotate, curate, and augment works of art within and beyond their respective collections [34]. *Curarium* is an application for exploring, analyzing, and making arguments about the works of art in art collections, and it allows users to annotate these works of art, tell stories about each work, and curate collections in a collaborative way. Similarly, the *Google Art Project* was created in 2011, currently running as Google Arts & Culture platform in collaboration with 151 museums from 40 countries. The online platform enables people to access high-resolution images of more than 32,000 artworks, allowing them to take virtual tours of partner museums galleries with the use of various multimedia features. Users can explore physical and contextual information about artworks and can compile and share their own virtual collections [26].

3.3. Summary

Based on our interview results, we identified three themes for opportunities in need of more attention if we want to improve participatory action, accessibility, and diversity in cultural heritage. These include (1) redefining the meaning of "expert" by including also local audiences, (2) extending the experience space beyond cultural heritage sites by allowing offsite discoveries as part of an onsite experience, and (3) providing an infrastructure to decentralise collaboration for cultural heritage. While the first was raised based on issues mentioned by general audience interviewees and relates to Kidd's marketing frame, the other two were based on the main challenges that cultural heritage professionals are currently facing and relate to Kidd's inclusive and collaborative frames [27].

4. Study 2: A Case Study

To further support findings from the in-depth interviews, we conducted a case study around built heritage using audio narratives. Heritage has a compelling role in postmodern tourism, specifically towards the built side of things. Architectural collections are popular tourist attractions, which according to Nuryanti can be defined as *the heart of cultural tourism* [35]. The relationship between tourism and built heritage, however, brings to frictions that usually emerge due to the juxtaposition between tradition and modernity. Built heritage can be used as an artefact to promote ethics, history, or industry but can also be experienced by visitors through a narrow understanding of time and place. Architecture can be an immersive space into the history of the city, but the general lack of knowledge towards architecture prevents people from engaging with the built environment. An intriguing perspective is that architecture could act as a common ground that is usually shared by a mixed public, including tourists and locals, while limited information is available to contextualise the artefacts. For this case study, we chose the square at Kensington Gore in central London, which contains such a built heritage, allowing views of historical buildings and monuments. The built heritage of interest in this square are the following:

- the Royal Albert Hall;
- the Prince Albert Sculpture, known as the "Great Exhibition Memorial";
- the Royal College of Music Building;
- the Albert Court; and
- the Beit Quadrangle.

This square is usually a space that both local audiences and tourists share during office breaks or while passing through. See Figure 1 for some views of the site.

Figure 1. Left image: view of the site with the Prince Albert sculpture and Albert Hall in the background. **Right image**: view of the site with the Royal College of Music Building and the Albert Court in the background.

The main objective of this study was twofold: first, to broaden the experience around the buildings and monument with public contribution and, second, to compare engagement between a traditionally/professionally curated audio narrative and a user-generated one.

The results of this study were expected to provide further details on implications on how existing social media could be better utilised to improve user engagement and to encourage distributed curation within a cultural heritage context, looking specifically at the challenges that this approach could generate.

4.1. Methodology

4.1.1. Audio Narratives

Firstly, for each built heritage listed in the previous section and shown in Figure 1, two types of audio narratives were created: a traditional curator narrative and a user-generated narrative. The curator narrative was created in partnership with an architectural historian who is trained in London's Victorian urban development. The user-generated narrative consisted of short excerpts inspired from public posts that were created during a co-design workshop. In this workshop, we explained to local audiences the concept of user-generated content. They visited the case study site and were invited to share their thoughts about the site on Instagram. To help them on the topic, we asked them the following questions:

- What does this building remind you of?
- Can you identify any specific artistic or architectural features?

Participants were asked to share their thoughts on Instagram using images and captions and were free to post as many or as little as they wanted. The public posts were then used to extract quotes for the user-generated narrative. Both curator narratives and user-generated narratives were kept short to avoid "museum fatique" [36]. See Appendix B for the transcripts of the narratives.

4.1.2. The 3D Tune-In Toolkit

Once recorded, all narratives were processed using the binaural spatialisation technique through the 3D Tune-in Toolkit [37] in order to create a sense of direction and space within the audio scenes. Binaural spatialisation is a technique that allows for the creation of three-dimensional soundscapes through a simple pair of headphones and is widely used in interactive audio and Virtual Reality (VR) applications. Recorded audio was processed using the 3D Tune-In Toolkit Test application (https://github.com/3DTune-In/3dti_AudioToolkit/releases, accessed on 30 January 2020), creating an immersive soundscape with sound sources located all around the listener at different distances. The spatialised stereo audiofiles were then exported and used for the test.

Recently, within the work of the *PLUGGY* project, a Javascript version of this platform was released called PlugSonic Soundscape, which allows us to perform the same process and to create immersive binaurally spatialised soundscapes using a simple web interface [38].

Immersive virtual environments have been around for several years [39] and have been extensively used in various research fields, both in audiovisual [40,41] and audio-only modes [42]. Previous research has indicated that immersive VR systems can stimulate the experience of spatial presence, albeit this might be mainly based on the visual communication channel [40]. This could be a key outcome considering that, in this study, we are exploring the role that immersive audio could play when looking at making cultural heritage more accessible.

4.1.3. Design Probe

The binaural audio narratives were made available through a simple website created using WordPress. The website included five separate pages for each building or monument; in each page, there was an audio file that could be played back when pressing a specific button on the interface. Each page also had a text version of the narrative. To allow participants to easily access the content, the audio narratives pages were also directly accessible through five QR codes printed and attached to the sides of a laser-cut cube that participants hold on to in their hands throughout the experience (see Figure 2). Each cube also had an image of the building or monument printed and attached to the cube so that participants could quickly find the audio narrative of interest. By scanning a QR code, participants were immediately directed to the correct web page. On one side of the cube, open questions were presented to trigger participants to comment on the narratives: *"What does this building remind you of?"* and *"What influences do you see?"*. Participants were invited to add their comments or pictures on social platforms such as Instagram using a specific hashtag or in the news and micro blogging platform *Medium* (www.medium.com, accessed on 30 January 2020) using specific tags.

Figure 2. Each participant used a cube with images and QR codes on the various sides to access the text and audio information for each target. On one side of the cube, questions were shown to trigger visitors thoughts when roaming around.

4.1.4. Participants

Through opportunity sampling, a total of 28 participants were recruited either through the local university network or were randomly approached on the spot while walking through Kensington Gore. Fifteen participants were given access only to the curator narrative. Thirteen participants were given access to both the curator and the user-generated narrative. The study was approved by the Imperial College Ethics commission.

4.1.5. Procedures

The study took place in Q2 of 2019. At the start of the study, participants were introduced to the cube and explained how to access the narratives. No specific instructions were given other than to freely select the audio narratives in the order they wanted. They were allowed to walk anywhere within the site. Participants were either given a smartphone and noise cancelling earphones, or they used their own devices to access the audio narratives. Since the site was not close to a main road and has generally not many passing visitors during the week, background noise was not considered an issue [43]. During their explorations, participants were shadowed and observed. A map of the test site was used to annotate navigation, touch points, and behavioural insights. Participants decided when they finished by reporting back to the observer. At the end of study, they were invited to answer a short version of the MEC-SPQ (Measurements, Effects, Conditions Spatial Presence Questionnaire) to measure engagement [44]. The MEC-SPQ was designed for immediate assignment after media exposure, and the short version consists of 28 question items representing seven constructs including process factors (attention allocation, spatial situation model, self location, and possible actions), variables referring to states and actions (higher cognitive involvement and suspension of disbelief), and variables addressing enduring personality factors (i.e., domain-specific interest). Each construct contains 4 questions that can be answered on a 5-point Likert scale ranging from 1 ("I do not agree at all") to 5 ("I fully agree"). Here, a short description of each constructs follows:

- attention allocation, i.e., to understand which media was more effective in communicating, written text, images, or audio;
- spatial situation model, i.e., to understand the users' ability to describe to what extent it was possible to link the content to their physical surroundings;
- self-location, i.e., to understand how grounded the users felt in relation with the mixed-media environment;
- possible actions, i.e., the sensation of being able to take action in the mediated environment;
- higher cognitive involvement, i.e., to understand how the content activated the thinking of the users;
- suspension of disbelief, i.e., to understand if the users were able to spot errors in the content or were biased toward the user-generated content; and
- domain-specific interest, i.e., to understand how much the users were interested in the topics related to the study.

The group of participants who had the option to listen to both the curator narrative and the user-curated narrative was asked to comment on and to explain their preferences. Demographics were also collected at the end.

4.1.6. Analysis

Descriptive statistics were used to analyse the MEC-SPQ scores for participants with access to the curator narrative vs. both curator and user-generated narratives. We also conducted a thematic analysis, similarly to what we did in the interview study but specifically focusing on participants demographics, and their decisions and comments with regard to their preferences. Again, pseudonyms are used to refer to participants.

4.2. Results

Overall, the curator narrative was found to be more engaging. Specifically, participants who experienced the curator narrative gave higher scores for self location, possible actions, higher cognitive involvement, and domain-specific interest (see Table 1). The scores differ from participants who experienced the user-generated content by at least 0.5. For the other constructs, no differences were found.

Table 1. Average MEC-SPQ (Measurements, Effects, Conditions Spatial Presence Questionnaire) scores for each construct for participants who experienced the curator narrative vs. those who experienced the user-generated narrative. In bold are the scores with at least a difference of 0.5 in comparison to the other participant group.

MEC Spatial Presence Item Category	Curator Narrative M(SD)	User-Generated Narrative M(SD)
Attention allocation	3.77(0.93)	3.83(0.68)
Spatial Situation Model	3.54(1.18)	3.60(1.18)
Spatial Presence: Self Location	**3.87(1.10)**	3.32(1.32)
Spatial Presence: Possible Actions	**3.79(1.28)**	3.00(1.17)
Higher Cognitive Involvement	**4.24(0.66)**	3.56(1.03)
Suspension of Disbelief	2.74(1.23)	2.74(1.16)
Domain Specific Interest	**3.77(1.25)**	3.23(1.11)

The Role of Familiarity

Based on the qualitative data (i.e., observations and user comments), participants showed preferences for the curator narrative because they trusted it more than the user-generated narrative. This was the case especially for participants who were less familiar with the context and who, therefore, wished to grasp most of the knowledge about the area from an expert. For example, Ella said *"I'm from Sydney, Australia, and I just arrived in the UK yesterday. I'm not very familiar with the surroundings and I'd like to know more about that statue (referring to the Albert Memorial) from a more official source."* On the other hand, participants who were more familiar with the surroundings preferred the user-generated narrative in order to check other people's experience about the same place, i.e., Paul said *"I live in London since six years now and I work nearby since four years. I think, letting people share their experience about the same places is very nice, especially because I'm always interested to get to know what other Londoners know about a specific area."*

The cultural background also seemed to play a key role in participants' preferred narrative type. It made a difference on the amount and the diversity of details the users mentioned to be able to grasp from the surroundings, which is supported by previous research [45]. For example, some participants were not always familiar with the Latin inscriptions under the Albert Memorial monument, and they expected a prompt translation from the audio narrative. This was the case for participants who came from eastern Asia and some European countries where Latin is not taught in secondary school. An audio narrative required a high attention span for non-English native speakers, especially if the content was delivered in a speech form. For this reason, participants in this specific case preferred to also read the text while listening to the audio narrative.

Only when participants were familiar with the heritage site (because of a previous experience related with the site or because of the cultural background) or have a similar cultural background did they prefer to listen to the user-generated narrative. Therefore, in order to consider local audiences as experts or locals as an important marketing channel, the cultural background and knowledge of the content consumers should be taken into consideration. This supports previous research [46]; user-generated content may only be interesting to those with familiarity of the context [47].

4.3. Summary

Based on the results of this second part of the study, the curator narrative was generally the preferred one. The only exceptions occurred when participants were already familiar with the site; in that case, the user-generated content was more appreciated. This seems to suggest that curator content should always be presented while user-generated stories can be offered as an additional experience depending on visitors familiarity with the artefact or environment.

5. Discussion and Conclusions

This work supports previous projects such as CURIOS [1], among others, that have shown the potential of social platforms to build networks through distributed contributions of users but within the context of cultural heritage. In this work, we identified issues and challenges in involving people in distributed curation in cultural heritage and explored a potential design space and opportunities for social media and audio technology to improve accessibility and awareness and to promote local heritage [23].

Our findings show that current social media could be further utilised to improve distributed curation, accessibility, and diversity in cultural heritage within 3 themes: firstly, that we should consider local audiences as experts and, therefore, as an important resource and potential marketing channel to become more social within the new museology [27]. However, as we have learned from the case study, we do need to take into account visitors' cultural background and level of familiarity. These seem to have some impact on visitors' preferences in curator vs. user-generated content. The latter may only be interesting to those familiar with the site. Hence, the type of content should be tailored not only to different cultural backgrounds but also to different knowledge levels, which depend on the audience rather than the cultural topic itself. This finding, however, may only be applicable to the type of built heritage used in this case study. We do have to note that this may be different for different types of sites, e.g., pleasant scientific sites vs. dark history sites.

We have also shown the need to extend the experience space by blending the discovery and experiences of heritage artefacts into everyday spaces. In the case study, we faced a major challenge in engaging visitors who were passing through. We believe the challenge may have been in the nature of the context as its part of participants' "everyday" or "work" environment and visitors were just passing through (i.e., a transitional place) e.g., to get to one of the museums nearby. More exploration should be performed on how cultural heritage can be introduced into people's "everyday" or "transitional" spaces, where relevant heritage cstories could be discovered that are related to other heritage sites or artefacts inside a museum nearby.

Additionally, there seems to be a need for an infrastructure to decentralise collaboration for cultural heritage, so that general audiences can be content creators anywhere and at anytime for objects onsite and/or offsite museums or heritage sites. In order to allow a general international audience to create content, technology should support easy access to curation tools that can be used by anyone and at anytime. The rising importance of digital distributed curation is important to improve inclusion and may even become more valued during current times such as during the COVID-19 pandemic [4,48].

Our work contains a few limitations. Although we highlighted a number of opportunities for further exploration, our findings do not offer a complete insight into potential opportunities in social media to better engage people at a global level. This is due to the qualitative nature of our studies. Generally, the sample size used in qualitative research methods is smaller than in quantitative research methods because qualitative research methods intend to gain in-depth understanding of a phenomenon and, as in this work, are often centered on the how and why of a particular issue. As described in [49], in-depth interview work is not as concerned with making generalisations to a larger population of interest and does not tend to rely on hypothesis testing but rather is more inductive and emergent in its process. Therefore, we selected interviewees based on expected types of initial users of PLUGGY. Another limitation is that, in addition to motivating visitors to participate on the spot, they may need to help support quality checks of distributed curation. This was not addressed in this work. Furthermore, it is also important to note that, when encountering foreign cultures, each has different ways to approach content. As we have seen, not only do peoples' cultures but also familiarity and previous experiences play significant role. A meaningful cultural experience, therefore, needs to be presented in different ways for different audience types.

To conclude, the work presented here could be of value in helping to identify opportunities in the use and design of social media and technology, which could further support participatory action and, hence, inclusivity. Inclusivity is a key to promote awareness. Additionally, wthout proper access to cultural content, people are unable to engage in a meaningful experience with cultural heritage. However, further work is necessary to explore different case studies within various types of transitional locations and media content types.

Author Contributions: Conceptualization, V.L., S.K. and L.P.; methodology, V.L. and S.K.; software, L.P.; validation, V.L. and S.K.; formal analysis, V.L. and S.K.; investigation, V.L. and S.K.; resources, V.L. and L.P.; data curation, V.L., S.K. and L.P.; writing—original draft preparation, V.L. and S.K.; writing—review and editing, V.L. and L.P.; visualization, V.L. and S.K.; supervision, L.P.; project administration, L.P.; funding acquisition, L.P. All authors have read and agreed to the published version of the manuscript.

Funding: *PLUGGY* received funding from the European Union's Horizon 2020 Research & Innovation Programme under grant agreement no 726765. The content reflects only the author's view, and the European Commission is not responsible for any use that may be made of the information it contains.

Institutional Review Board Statement: The study was conducted according to the guidelines of the Declaration of Helsinki, and approved by the Imperial College Research Ethics Committee (reference: 17IC4197; approved in October 2017).

Informed Consent Statement: Written informed consent was obtained from the patient(s) to publish this paper.

Acknowledgments: We greatly thank our project partners Luca Simeone, Spyros Souvlakis, Luis Molina-Tanco, and Silvia Brandelesi for helping with the transcriptions and providing feedback on initial drafts of the paper. This work is a part of the *PLUGGY* project.

Conflicts of Interest: The authors declare no conflict of interest. The funders had no role in the design of the study; in the collection, analyses, or interpretation of data; in the writing of the manuscript; or in the decision to publish the results.

Appendix A

Appendix A.1. Questions Asked to Cultural Heritage Professionals

1. Tell me a bit about yourself. What you do on a daily basis?
2. What projects have you been working on or will be working on?
3. What is the process of content creation up to delivery to general audience?
4. What was the role of social media?
5. What were the main goals to achieve?
6. What were your main challenges?
7. What it is that made something (e.g., an aspect of technology) successful or not?
8. What do your audience want and like(consider different segmentation of audience)?
9. Have you observed any shifts of use or behaviours among your audience?
10. What would you like to do in the future that you cannot do right now?
11. How do you think about crowdsourced content? What challenges do you foresee?

Appendix A.2. Questions Asked to a General Audience

1. Tell me a bit about yourself. What do you do on a daily basis?
2. Can you tell me about cultural heritage in your home city?
3. Have you been to any cultural heritage related sites or events recently? Tell me about it?
4. What did you do in preparation of the visit?
5. What were your main motivations and goal of the visit?
6. When and where did you get information?

Appendix B

Appendix B.1. Curator Narratives

Appendix B.1.1. Royal Albert Hall

"It's beautiful isn't it? The Royal Albert Hall. It seems like only yesterday when I was talking to Henry Cole, chairman of the royal society of arts, about my dream to build a permanent exhibition hall. A sort of 'follow up' of the Great Exhibition of 1851. Oh, by the way, my name is Albert... and I was the Prince of Saxe-Coburg and Gotha. I had a beautiful wife, Victoria, queen of England. If it wasn't for her and Henry, the royal albert hall wouldn't exist. The Hall was opened on 29 March 1871 by Queen Victoria, renamed in my memory to the Royal Albert Hall of Arts and Sciences. When Victoria opened the Hall, she was so overcome by emotion that the Prince of Wales had to speak in her place... She was one of the most beautiful women on this world. Unfortunately, I wasn't there at the inauguration. But they told me that as she laid the foundation stone, she said 'It is my wish that this hall should bear his name to whom it will have owed its existence and be called The Royal Albert Hall of Arts and Sciences'. But the Royal Albert Hall is part of my dream resulted in the construction of Albertopolis. The area is here in South Kensington and contains several structures dedicated to the arts and sciences. For example, I'm now facing the Imperial College London. Ah. the Imperial College... the name me and Victoria chose for the building was

Imperial Institute! Unfortunately, I didn't have time to see the opening of this building either... due to my early death. The college was one of the first in Britain to teach by experiments rather than just by lectures."

Appendix B.1.2. Royal College of Music

"Within sight of the Royal Albert Hall are two buildings dedicated to music. The first one, with your back to the Hall, is the Royal College of Music. It originated from my proposals for a national music training scheme for young people. Founded in 1882, it is now part of the University of London. The Royal College of Music is one of the world's leading conservatoires. It provides specialised musical education and professional training at the highest level for performers, conductors and composers. Some of the world's best classical musicians have studied here. Listen carefully and you may be able to hear students playing inside."

Appendix B.1.3. Albert Court

"The Albert Court is one of the first historical Flat housing facilities in the UK. This mansion block was originally conceived by the freeholders and commissioners, as the first stage of a larger private development. Building began in 1890 to the designs of Frederick Hemings but, following the collapse of the Liberator Building Society in 1892 and the death of Hemings in 1894, the building had only reached the 3rd floor level. The Albert Court was finished by R.J. Worley's designs between 1896 and 1900. The time it took to complete this building reminds me the troubled and long path that the Royal Albert Hall had to go in order to become the magic building the whole world envies to Britain. This building is made with red brick with elaborated stonework bands and dressings. It has 6 storeys, a D-shaped plan, with the North and East facades following the crescent line of the former Royal Horticultural Society's Garden."

Appendix B.1.4. Beit Hall

"Beit Hall, otherwise known as the Beit Quadrangle, is one of Imperial College London's oldest and most historic buildings. It was funded by Sir Otto Beit in 1910, then director of the governing body of Imperial College. Initially it was a faculty building housing Biology and later converted to a student Hall. The band Queen is said to have performed here its first gig. Their music was definitely different to what I was used in the 19th Century. During term-time, it is one of the largest Halls of Imperial College providing self-catering accommodation to 340 students. But during the remaining 14 weeks Beit magically transforms into an international conference centre and hotel. The colour that architect Aston Webb chose for the bricks perfectly matches the orange of my beloved royal albert hall."

Appendix B.1.5. Albert Memorial Statue

"Look at me! This statue, which represents me, is a memorial to the Great Exhibition. Until 1891 it stood in the garden of the Royal Horticultural Society, which used to be on the site of the Royal College of Music. Take a look at the inscription below this statue, which sets out how much money we raised from the Great Exhibition. Albertopolis actually all started with the Great Exhibition in 1851, which showcased Britain's international role in the arts and sciences. It was the first of a series of world fairs for culture and industry, where scientists used to meet, discuss and share ideas and innovations. From the Exhibition's success this area south of Hyde Park was established as a long-term legacy to celebrate science, technology, culture and the arts. We have seen how Albertopolis is home to some of the world's leading museums, academic institutions and national organisations. Each one of these is continually evolving and expanding, through their buildings and the people who work in, study in and visit them. More than 150 years later Prince Albert and Sir Henry Cole's legacy is still alive and well. Albertopolis is still at the heart of the arts and sciences."

Appendix B.2. User-Generated Audio Narrative

Appendix B.2.1. Royal Albert Hall:

Excerpt 1: *"These decorations you see under the roof remind me of pottery from ancient Greece"*—sound location: top right.

Excerpt 2: *"The reliefs remind me of the Greek reliefs in the British Museum taken from the Parthenon. They are definitely different but somehow I like the idea of relief representation that it comes from the Greek influence?"*—sound location: top left.

Excerpt 3: *"I used to dream of being a musician. Whenever I see the building, I wonder what it would have been like to be a musician, and I feel a bit of pity for my lost dream."*—sound location: middle front.

Excerpt 4: *"The architecture of the façade is surprisingly similar to the San Felice Sul Panaro church near Modena, which sadly has been destroyed in the recent earthquake."*—sound location: middle front.

Appendix B.2.2. Royal College of Music:

Excerpt 1: *"On both side of the main entrance there are musical themed reliefs, and in the spandrels above them are carved wreath motifs."*

Excerpt 2: *"You can see the pilasters travel up the building and change into two small niches, which then move upwards and change into finials on the central dormer. This has a clock mounted in the top section with a triangular sun relief and strap work decorations."*

Appendix B.2.3. Albert Memorial Statue:

Excerpt 1: *"I'm curious about the 4 statues placed in the base of the Albert memorial. Their attire looks different from each other, maybe they symbolise the 4 corners of the world?"*

Excerpt 2: *"In the other Albert Memorial in Hyde park there are some Marble figures and animals representing Europe, Asia, Africa and America at each corner. For the animals, Africa is represented by a camel, the Americas with a buffalo, Asia by an elephant and a bull for Europe."*

References

1. Webster, G.; Nguyen, H.; Beel, D.E.; Mellish, C.; Wallace, C.D.; Pan, J. CURIOS: Connecting Community Heritage through Linked Data. In Proceedings of the 18th ACM Conference on Computer Supported Cooperative Work (CSCW'15), Vancouver, BC, Canada, 14–18 March 2015; ACM: New York, NY, USA, 2015; pp. 639–648.
2. Charles, A. What is the 1% rule? *The Guardian*, 20 June 2006.
3. Lim, V.; Frangakis, N.; Tanco, L.M.; Picinali, L. PLUGGY: A Pluggable Social Platform for Cultural Heritage Awareness and Participation. In *Advances in Digital Cultural Heritage*; Ioannides, M., Martins, J., Žarnić, R., Lim, V., Eds.; Springer International Publishing: Cham, Switzerland, 2018; pp. 117–129.
4. Fairclough, G.; Šešić, M.D.; Mijatović, L.R.; Auclair, E.; Soini, K. The faro convention, a new paradigm for socially-and culturally-sustainable heritage action. *Culture* **2015**, *8*, 9–19.
5. Council of Europe Treaty Series-No. 199. 2005. Available online: https://rm.coe.int/16800d3814 (accessed on 1 June 2017).
6. Murzyn-Kupisz, M.; Działek, J. Cultural heritage in building and enhancing social capital. *J. Cult. Herit. Manag. Sustain. Dev.* **2013**, *3*, 35–54. [CrossRef]
7. Vinson, I. ICCROM, Äôs contribution to the ethics of heritage. *Mus. Int.* **2009**, *61*, 90–97. [CrossRef]
8. Yoshimura, Y.; Sobolevsky, S.; Ratti, C.; Girardin, F.; Carrascal, J.P.; Blat, J.; Sinatra, R. An Analysis of Visitors' Behavior in the Louvre Museum: A Study Using Bluetooth Data. *Environ. Plan. Plan. Des.* **2014**, *41*, 1113–1131. [CrossRef]
9. Vivant, E. Who brands whom? The role of local authorities in the branching of art museums. *Town Plan. Rev.* **2011**, *82*, 99–115. [CrossRef]
10. Wolferstan, S.; Fairclough, G. Common European heritage: Reinventing identity through landscape and heritage? *Fram. Eur. Perspect. Eur. Archaeol. Counc. Pap.* **2013**, *7*, 43–54.
11. Giaccardi, E.; Churchill, E.; Liu, S. Heritage Matters: Designing for Current and Future Values through Digital and Social Technologies. In Proceedings of the CHI '12 Extended Abstracts on Human Factors in Computing Systems, Austin, TX, USA, 5–10 May 2012; ACM: New York, NY, USA, 2012; pp. 2783–2786. [CrossRef]
12. Cushing, B. Always connected beyond broadband-what comes next? In *Broadband Services, Applications, and Networks: Enabling Technologies and Business Models*; International Engineering Consortium. 2011; pp. 15–23. Available online: https://shop.brainlairbooks.com/book/9781931695244 (accessed on 26 March 2021).
13. Howe, J. The rise of crowdsourcing. *Wired Mag.* **2006**, *14*, 1–4.

14. Liu, S.B. The Living Heritage of Historic Crises: Curating the Bhopal Disaster in the Social Media Landscape. *Interactions* **2012**, *19*, 20–24. [CrossRef]
15. Weilenmann, A.; Hillman, T.; Jungselius, B. Instagram at the Museum: Communicating the Museum Experience Through Social Photo Sharing. In Proceedings of the SIGCHI Conference on Human Factors in Computing Systems (CHI'13), Paris, France, 27 April–2 May 2013; ACM: New York, NY, USA, 2013; pp. 1843–1852.
16. Yelmi, P.; Kuşcu, H.; Yantaç, A.E. Towards a Sustainable Crowdsourced Sound Heritage Archive by Public Participation: The Soundsslike Project. In Proceedings of the 9th Nordic Conference on Human-Computer Interaction (NordiCHI '16), Gothenburg, Sweden, 23–27 October 2016; ACM: New York, NY, USA, 2016; pp. 71:1–71:9. [CrossRef]
17. Cosley, D.; Baxter, J.; Lee, S.; Alson, B.; Nomura, S.; Adams, P.; Sarabu, C.; Gay, G. A Tag in the Hand: Supporting Semantic, Social, and Spatial Navigation in Museums. In Proceedings of the SIGCHI Conference on Human Factors in Computing Systems (CHI '09), Boston, MA, USA, 4–9 April 2009; ACM: New York, NY, USA, 2009; pp. 1953–1962.
18. Han, K.; Shih, P.C.; Rosson, M.B.; Carroll, J.M. Enhancing Community Awareness of and Participation in Local Heritage with a Mobile Application. In Proceedings of the 17th ACM Conference on Computer Supported Cooperative Work (CSCW '14), Baltimore, MD, USA, 15–19 February 2014; ACM: New York, NY, USA, 2014; pp. 1144–1155.
19. McGookin, D.; Brewster, S. PULSE: An auditory display to provide a social vibe. In Proceedings of the Interacting with Sound Workshop: Exploring Context-Aware, Local and Social Audio Applications (IwS '11), Stockholm, Sweden, 30 August–2 September 2011; ACM: New York, NY, USA, 2011; pp. 12–15.
20. Bellens, A.; Banc, N.V.L.; Eloire, F.; Grabar, N.; Kergosien, E.; Severo, M. Social Media and European Cultural Routes: Instagram Networks on the via Francigena. In Proceedings of the 8th International Conference on Management of Digital EcoSystems (MEDES), Biarritz, France, 1–4 November 2016; ACM: New York, NY, USA, 2016; pp. 122–128. [CrossRef]
21. Bujari, A.; Ciman, M.; Gaggi, O.; Palazzi, C.E. Using Gamification to Discover Cultural Heritage Locations from Geo-tagged Photos. *Pers. Ubiquitous Comput.* **2017**, *21*, 235–252. [CrossRef]
22. Srivastava, S. A Study of Awareness of Cultural Heritage among the Teachers at University Level. *Univers. J. Educ. Res.* **2015**, *3*, 336–344. [CrossRef]
23. Nyaupane, G.P.; Timothy, D.J. Heritage awareness and appreciation among community residents: Perspectives from Arizona, USA. *Int. J. Herit. Stud.* **2010**, *16*, 225–239. [CrossRef]
24. Nasser, N. Planning for urban heritage places: Reconciling conservation, tourism, and sustainable development. *J. Plan. Lit.* **2003**, *17*, 467–479. [CrossRef]
25. Hiller, H.H. Conventions as mega-events: A new model for convention-host city relationships. *Tour. Manag.* **1995**, *16*, 375–379. [CrossRef]
26. Ardissono, L.; Kuflik, T.; Petrelli, D. Personalization in cultural heritage: The road travelled and the one ahead. *User Model.-User-Adapt. Interact.* **2012**, *22*, 73–99. [CrossRef]
27. Kidd, J. Enacting engagement online: Framing social media use for the museum. *IT People* **2011**, *24*, 64–77. [CrossRef]
28. Bozdag, E. Bias in Algorithmic Filtering and Personalization. *Ethics Inf. Technol.* **2013**, *15*, 209–227. [CrossRef]
29. Alexander, C.; Burnette, A.; Dark, D.; Hart, D.; Rossi, J.; Minor, N. Beyond Launch: Museum Videos on YouTube. 2008. Available online: http://www.archimuse.com/mw2008/papers/hart/hart.html (accessed on 12 May 2019).
30. Bernstein, M.; Bolter, J.D.; Joyce, M.; Mylonas, E. Architectures for Volatile Hypertext. In Proceedings of the Third Annual ACM Conference on Hypertext (HYPERTEXT '91), San Antonio, TX, USA, December 1991; ACM: New York, NY, USA; pp. 243–260. [CrossRef]
31. Roberts, R.C. Questions of Museum Essence: Being, Being With, and Finding Connection in Conversation. *Museums Soc. Issues* **2013**, *8*, 89–101. [CrossRef]
32. Sachatello-Sawyer, B.; Burton, H. *Adult Museum Programs: Designing Meaningful Experiences*; Rowman Altamira: Lanham, MD, USA, 2002.
33. Battarjee, K.; Koskinen, I. Co-experience: User experience as interaction. *CoDesign* **2005**, *1*, 5–18. [CrossRef]
34. Simeone, L.; Secundo, G.; Schiuma, G. Arts and design as translational mechanisms for academic entrepreneurship: The metaLAB at Harvard case study. *J. Bus. Res.* **2018**, *85*, 434–443. [CrossRef]
35. Nuryanti, W. Heritage and postmodern tourism. *Ann. Tour. Res.* **1996**, *23*, 249–260. [CrossRef]
36. Bitgood, S. Museum fatigue: A critical review. *Visit. Stud.* **2009**, *12*, 93–111. [CrossRef]
37. Cuevas-Rodrguez, M.; Picinali, L.; Gonzlez-Toledo, D.; Garre, C.; de la Rubia-Cuestas, E.; Molina-Tanco, L.; Reyes-Lecuona, A. 3D Tune-In Toolkit: An open-source library for real-time binaural spatialisation. *PLoS ONE* **2019**, *14*, e0211899. [CrossRef]
38. Comunità, M.; Gerino, A.; Lim, V.; Picinali, L. Design and Evaluation of a Web-and Mobile-Based Binaural Audio Platform for Cultural Heritage. *Appl. Sci.* **2021**, *11*, 1540. [CrossRef]
39. Slater, M.; Usoh, M. Representations systems, perceptual position, and presence in immersive virtual environments. *Presence Teleoper. Virtual Environ.* **1993**, *2*, 221–233. [CrossRef]
40. Edler, D.; Kühne, O.; Keil, J.; Dickmann, F. Audiovisual cartography: Established and new multimedia approaches to represent soundscapes. *KN-J. Cartogr. Geogr. Inf.* **2019**, *69*, 5–17. [CrossRef]
41. Indans, R.; Hauthal, E.; Burghardt, D. Towards an Audio-Locative Mobile Application for Immersive Storytelling. *J. Cartogr. Geogr. Inf.* **2019**, *69*, 41–50. [CrossRef]

42. Picinali, L.; Afonso, A.; Denis, M.; Katz, B.F. Exploration of architectural spaces by blind people using auditory virtual reality for the construction of spatial knowledge. *Int. J. -Hum.-Comput. Stud.* **2014**, *72*, 393–407. [CrossRef]
43. D'Orazio, D.; Montoschi, F.; Garai, M. Acoustic comfort in highly attended museums: A dynamical model. *Build. Environ.* **2020**, *183*, 107176. [CrossRef]
44. Vorderer, P. MEC Spatial Presence Questionnaire (MECSPQ): Short Documentation and Instructions for Application. *Report to the European Community, Project Presence: MEC (IST-2001-37661)*. 2004. Available online: http://www.ijk.hmt-hannover.de/presence (accessed on 10 October 2018).
45. Deaux, K. Reconstructing social identity. *Personal. Soc. Psychol. Bull.* **1993**, *19*, 4–12. [CrossRef]
46. Liu, X. On Locality and Public Ownership of Intangible Cultural Heritage. *J. Guangxi Univ. Natl. Philos. Soc. Sci. Ed.* **2008**, *3*.
47. Ayeh, J.K.; Au, N.; Law, R. "Do we believe in TripAdvisor?" Examining credibility perceptions and online travelers' attitude toward using user-generated content. *J. Travel Res.* **2013**, *52*, 437–452. [CrossRef]
48. Samaroudi, M.; Rodriguez Echavarria, K.; Perry, L. Heritage in lockdown: Digital provision of memory institutions in the UK and US of America during the COVID-19 pandemic. *Mus. Manag. Curatorship* **2020**, *35*, 337–361. [CrossRef]
49. Dworkin, S.L. Sample Size Policy for Qualitative Studies Using In-Depth Interviews. *Arch. Sex. Behav.* **2012**, *41*, 1319–1320. [CrossRef] [PubMed]

applied sciences

MDPI

Article

Automated Graffiti Detection: A Novel Approach to Maintaining Historical Architecture in Communities

Jongseong Choi [1,2,*], Lazaros Toumanidis [3], Chul Min Yeum [4], Patrikakis Charalampos [3], Ali Lenjani [5], Xiaoyu Liu [6], Panagiotis Kasnesis [3], Ricardo Ortiz [1], Ning-Jun Jiang [7] and Shirley J. Dyke [6,8]

1 Department of Mechanical Engineering, SUNY Korea, State University of New York, Incheon 21985, Korea; ricardo.ortiz@stonybrook.edu
2 Department of Mechanical Engineering, Stony Brook University, State University of New York, Stony Brook, NY 11794, USA
3 Department of Electrical and Electronics Engineering, University of West Attica, 12241 Egaleo, Greece; laztoum@uniwa.gr (L.T.); bpatr@uniwa.gr (P.C.); pkasnesis@uniwa.gr (P.K.)
4 Department of Civil and Environmental Engineering, University of Waterloo, Waterloo, ON N2L 2G1, Canada; cmyeum@uwaterloo.ca
5 School of Medicine, Stanford University, Stanford, CA 94305, USA; alenjani@stanford.edu
6 School of Mechanical Engineering, Purdue University, West Lafayette, IN 47906, USA; liu1787@purdue.edu (X.L.); sdyke@purdue.edu (S.J.D.)
7 Institute of Geotechnical Engineering, Southeast University, Nanjing 211189, China; jiangn@seu.edu.cn
8 Lyles School of Civil Engineering, Purdue University, West Lafayette, IN 47906, USA
* Correspondence: jongseong.choi@stonybrook.edu; Tel.: +82-32-626-1811

Abstract: Graffiti is common in many communities and even affects our historical and heritage structures. This leads to a decrease in the revenue associated with commercial activities or services (e.g., shops, restaurants, residences), and potentially reduces tourism in a region. Visual data, in the form of photographs, is becoming an efficient mechanism to record information. Photographs can be quickly captured, and are already frequently posted online by ordinary citizens (e.g., tourists, residents, visitors). Exploiting image data through automation and computer vision provides a new opportunity to simplify the current manual graffiti-monitoring processes, enabling automated detection, localization, and quantification of such markings. In this study, we developed a vision-based graffiti-detection technique using a convolutional neural network. Images collected from historical structures of interest within a community can be utilized to automatically inspect for graffiti markings. In the case in which citizens collect and contribute data, there is a high degree of duplication and repetition, and potentially a lack of GPS information. These hinder the direct use of the images for automating the process. To address these challenges, we built high-resolution, single-view façade images (orthophotos) before applying our robust graffiti detector. The robust graffiti detector was built using a database with 1022 images of damaged or contaminated structures gathered during a recent European Union project, entitled "Safeguarding Cultural Heritage through Technical and Organisational Resources Management" (STORM). A total of 818 images were used for training (10% of the training set was randomly chosen for the validation set), achieving 88% accuracy among the remaining 204 samples for testing. Using the trained detector, the technique developed was demonstrated using data collected from the Church of Agios Nikolaos (Leontariou), Kantza, Greece.

Keywords: graffiti; cultural heritage assessment; convolutional neural network; orthophoto generation; object detection; citizen science

Citation: Choi, J.; Toumanidis, L.; Yeum, C.M.; Charalampos, P.; Lenjani, A.; Liu, X.; Kasnesis, P.; Ortiz, R.; Jiang, N.-J.; Dyke, S.J. Automated Graffiti Detection: A Novel Approach to Maintaining Historical Architecture in Communities. *Appl. Sci.* **2022**, *12*, 2983. https://doi.org/10.3390/app12062983

Academic Editors: Marek Milosz and Jacek Kęsik

Received: 30 January 2022
Accepted: 10 March 2022
Published: 15 March 2022

1. Introduction

The term *graffiti* is defined as "writing or drawings scribbled, scratched, or sprayed illicitly on a wall or other surface in a public place" [1]. Such markings are quite often understood as a manifestation of antisocial behavior performed to gain attention or as a form of thrill-seeking. In many cases, such markings are linked to vandalism, or even criminal behavior. A more devious form of graffiti, known as *tagging*, which refers to "the repeated use of a single symbol or series of symbols to mark territory", has become common in many places in the world. Numerous cases of graffiti-tinged vandalism have been documented on historical structures such as monuments, statues, churches, or temples. Figure 1 provides a sample of such vandalism. A Holocaust monument in central Athens was vandalized with tagging in December 2017, and in response, city officials stated that, "The attack on the monument, which will remain open and accessible to citizens, is an act of intolerance and historical ignorance" [2].

Figure 1. Vandalism of the Holocaust monument in Athens, Greece, in December 2017.

Graffiti causes various problems in a community [3–5]. First, there is the obvious cost to a community related to cleaning and maintenance, which can be significant, especially when graffiti has been applied to materials that require special treatment. Next, the presence of graffiti on walls and structures creates an impression of uncleanness and being unattended, and is thus often perceived as being linked to lack of surveillance or policing of the area in which the graffiti appears. This creates a sense of insecurity, and potentially lowers the value of land and property, and can lead to an overall decline in commercial activity and services (e.g., shops, restaurants, apartments). Furthermore, especially in the case of monuments and archaeological sites, graffiti can prove to be destructive. When the graffiti is sprayed or painted on older, delicate materials, cleaning requires special treatment, and the use of chemicals for the removal of graffiti may cause even more damage.

Current practice to locate and inspect graffiti takes the form of visual surveillance and evaluation by cleanup crews. This task is very time-consuming and requires significant resources [6–10]. The occurrence of such damage is quite unpredictable, so frequent visits may be needed. There is a compelling need to streamline this process by exploiting technologies and automation. In recent years, computer-vision technologies have begun to transform how structures are being inspected [11–19]. Deep-learning algorithms have drawn the most attention in both academia and industry because they have proven their ability to identify and localize visual features in images [20–25]. Image sensors are becoming smaller and cheaper, and nearly every ordinary citizen carries a cell phone on a daily basis. Thus, a large volume of high-quality visual data can be collected quite easily. By harnessing these capabilities, graffiti markings may be detected and monitored automatically, greatly reducing the costs associated with manual inspections.

However, there are several technical and social challenges that must be overcome to enable this sort of implementation. First, estimates of the locations and sizes of graffiti markings should be outcomes of the technique. Some images often capture a limited area without having sufficient spatial context. Thus, the markings detected are not correctly localized. Second, for the monitoring perspective, a sufficient quantity of data must be provided to gather up-to-date visual information of the community being assessed. Public crowdsourcing databases collected by many people (not only by certain engineers) might be utilized to gather the necessary data. Third, a robust graffiti detector should be built to accurately detect and localize graffiti markings in general situations from images. To ensure the robustness of the detector, a large database of ground-truth images is required.

The major innovation of this work is the preprocessing of orthophoto generation, merging individual citizen's image data in its workflow as a preprocessing. Graffiti markings are generally on long or large-size walls that cannot be captured in a single image; the orthophoto will improve the detection performance and enable quantification on large-scale structures. The capability of the system to estimate the sizes and locations of graffiti autonomously could help in remotely determining how much time and money should be invested in cleaning a specific section. On the other hand, the system allows the incorporation of images taken from ordinary citizens (people without the intention of doing visual inspection), which helps update the system without the requirement of scheduling visits to the site by experts or community staffs.

In this study, we developed a vision-based method to detect and quantify graffiti damage using images. We leveraged a ground-truth graffiti database generated by a recent European Union H2020 project, entitled "Safeguarding Cultural Heritage through Technical and Organisational Resources Management" (STORM). STORM has gathered a large volume of images of damaged or contaminated structures to be used for studying the community's long-term preservation strategies. Labeled graffiti images were utilized to build a robust graffiti detector. A key idea in the proposed technique was to generate a high-resolution orthophoto before the graffiti detector. To collect images periodically, various crowdsourcing strategies were used, such as a game applications on smartphones, harvesting from social media, promotions, advertisements to engage tourists, commuters, residents, and any other type of citizen. Once images over target structures in a community (TOCs) were collected, each façade was automatically estimated from the full set of images, yielding the appropriate locations and number façades within the TOC. An orthophoto for each façade of the structure was then generated by geometrically stitching multiple images collected from the TOC using a structure-from-motion (SfM) algorithm [26–30]. Then, the graffiti detector was applied to each orthophoto, which provided the relative size and location of each graffiti marking on the corresponding façade, thus localizing and quantifying the damage. All of these processes, including orthophoto generation and graffiti detection, were automated once we built the detector in advance. The capability of the technique was demonstrated using a cultural heritage site, the Church of Agios Nikolaos (Leontariou), Kantza, Greece [31].

2. Materials and Methods

The objective was to develop a technique that could detect graffiti using images gathered by citizens. To enable automation, an orthophoto was generated at various intervals to aid in decision making and planning regarding community maintenance needs. By allowing the citizens to collect the data, key structures in the community could be readily and frequently monitored against graffiti markings. A step-by-step procedure is depicted in Figure 2. As a preliminary step, a robust graffiti detector was trained. In this study, we used a robust convolutional neural network (CNN)-based object detector, called Faster R-CNN, to train the graffiti detector [32]. A real-world database with various graffiti scenes was used to train the robust detector [33]. Actual steps of the proposed technique were as follows. In Step 1, visual data were collected periodically from key structures that needed to be monitored. Ordinary citizens (e.g., residents, tourists, and visitors) were engaged to collect photographs to document relevant visual scenes in the community. The recommended guidelines for data collection are outlined in Section 2.2. In Step 2, orthophotos were generated that contained full views of façades in a target structure. An orthophoto is a high-resolution image, that is generated using the data collected in Step 1. The subject of this orthophoto, denoted here as the target object in a community (TOC), was represented by several orthophotos that each maintained their original image quality, sufficient for observing graffiti markings. Such orthophoto generation was a key step to enable image-based graffiti monitoring by removing any repetition or overlap that may have existed in the individual images, and then the TOC was represented by a single view per its façade. In Step 3, graffiti detection was performed by applying the trained graffiti detector to the reconstructed orthophotos. Since this process was repeated on the orthophotos reconstructed on a regular basis, such digital documentation was useful for instantly assessing the overall presence of graffiti on the TOC over time.

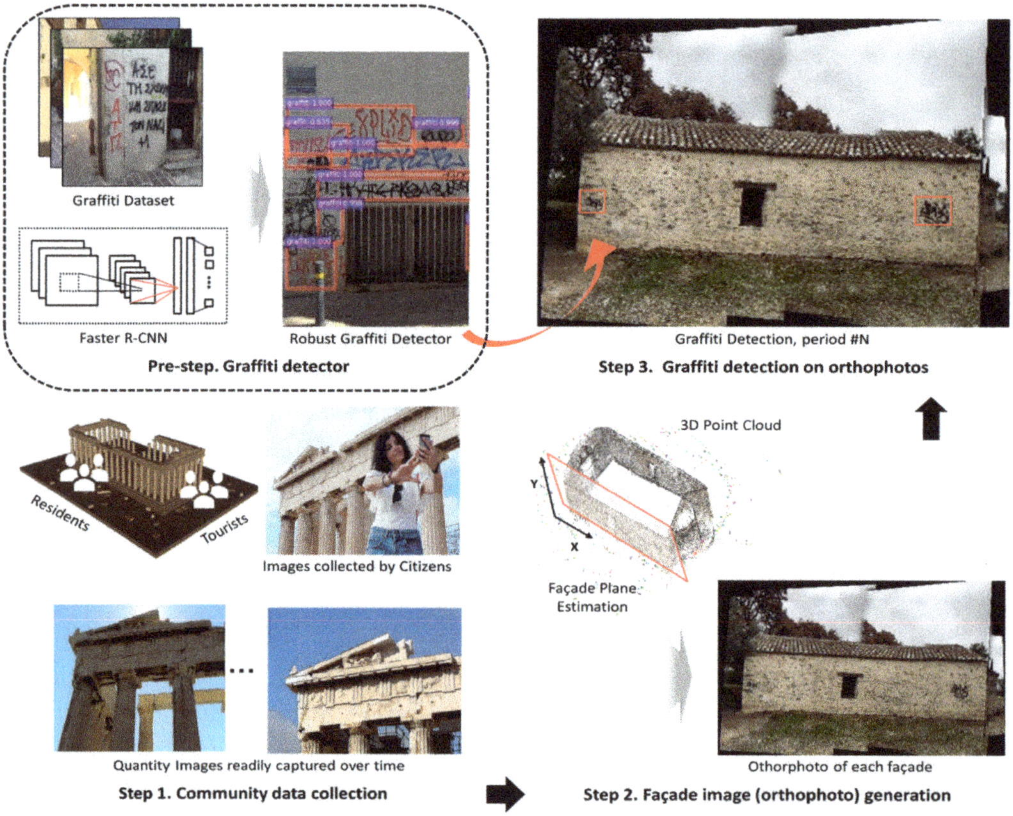

Figure 2. Overview of the technical approach. Prestep: develop a robust graffiti detector to exploit a graffiti database from STORM using Faster R-CNN; Step 1: collect images of target objects in a community (TOCs) periodically from ordinary citizens; Step 2: generate an orthophoto of each façade of a TOC by stitching and blending the collected images; Step 3: graffiti detection on orthophotos, including location and size of each graffiti.

2.1. Preliminary Step: Graffiti Detector

Graffiti is certainly not standardized in shape, color, or size, and thus must be distinguished from intentional signage that may exist in the background scene of an image.

Region-based convolution neural network (R-CNN) provides an appropriate solution to this problem. Recent advances in vision-based object detection are driven by the successful implementation of R-CNN (for example, see [34–36]). For building a graffiti detector, the basic requirement to use R-CNN, like all supervised learning procedures, is to retain a suitable training data set with a rich and diverse set of images captured from actual graffiti. The STORM project funded by the European Union (EU) provided a suitable data set, and we used it here for building a robust graffiti detector.

During the training phase of R-CNN, a step is required to extract candidate object areas, called a region proposal. CNNs operate by representing various types of graffiti scenes present in the set of region proposals with a unique feature vector. These region proposals may vary in size and scale, and therefore need to be reshaped to match the CNN input size. After reshaping, the region proposals pass through the CNN in order to generate the associated feature vectors. Then, these feature vectors are fed into the support vector machine (SVM) classifier. Lastly, the location and shape of each bounding box are refined by using a linear regressor to reduce the localization error [34]. R-CNN has evolved into a much faster algorithm called Fast R-CNN, which achieves greater efficiency by sharing expensive computations through a region-of-interest (ROI) pooling technique [35]. R-CNN has two major drawbacks that increase its computational cost. First, it requires a separate forward pass through the CNN to check if each region proposal includes an object class of interest. Then, it trains three different internal models in order: feature extraction, object classification, and bounding box regression.

To address these issues, Fast R-CNN generates a general feature map (for an entire region of each image) by passing each image (not its subregion) through the CNN and shares it with the ROI pooling layer to reduce the computational time. To find the max-pooled ROI proposals, the corresponding region of object proposals (likely positions of objects on the image) on the feature map should be extracted through max-pooling. Then, the extracted regions warp into a fixed-size feature vector and are passed to the fully connected layers. In the last layer of Fast R-CNN, the SVM classifier in R-CNN is replaced with a softmax layer for object classification [37]. A linear regression layer parallel to the softmax layer is also added to localize the object by outputting the coordinates of the bounding box. In this way, Fast R-CNN can integrate all necessary stages for training CNN (CNN feature extraction, object classifier, and bounding box coordinate regressor).

To develop a robust graffiti detector, we incorporated a state-of-art object detection method called a faster region-based convolutional neural network (Faster-RCNN), which is an enhanced version of Fast R-CNN in terms of computational expense. Figure 3 briefly portrays this process. In Faster R-CNN, the selective search algorithm was modeled with a fully convolutional network, called a region proposal network (RPN) [31]. RPN in Faster R-CNN made the generation of region proposals costless. RPN allowed features generated in the forward pass of CNN to be reused to produce the region proposals as well. Instead of implementing a separate proposal generator, RPN used an original image as the input, predicted region proposals, and fed those region proposals into the region-of-interest (ROI) layer for prediction of bounding boxes. RPN, which is a fully convolutional network, used the existing R-CNN to extract a feature map (the last output layer of R-CNN) for generating object proposals. After generating the feature map, RPN slid another convolutional layer on the feature map to produce the object proposals. The size of the sliding window that was spatially running on the feature map was defined based on the image size and aspect ratio of associated anchors. The anchors were fixed-size bounding boxes constructed with a combination of their predefined size and aspect ratios. RPN used anchors to produce a set of proposals by computing the probability that an anchor included an object (called the "objectness score") and adjusted the corresponding bounding box coordinates. Finally, the region proposals (outputs of the RPN) and feature maps (outputs of the shared CNNs) were used to train the Faster R-CNN to detect graffiti markings in each region proposal. Many other types of architecture were introduced to reduce the training and testing speed, as well as the accuracy, but there is always a trade-off between the accuracy and computational efficiency [38,39]. In this work, we implemented the original Faster R-CNN to detect graffiti markings on the images.

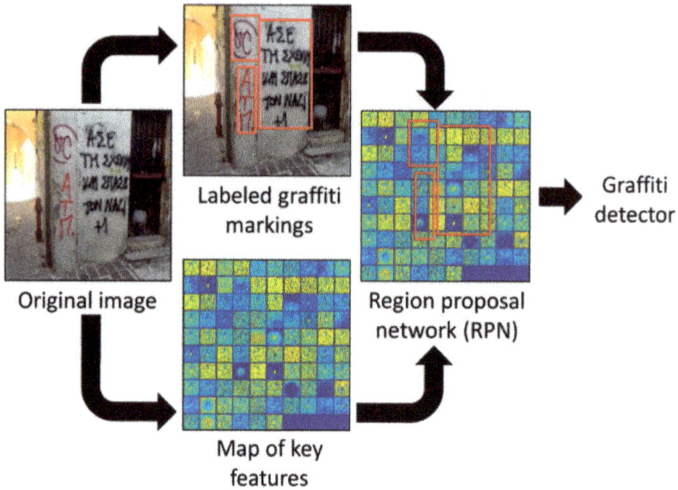

Figure 3. Faster R-CNN is a unified network of region proposals and key feature maps. This algorithm was used in this work to construct a robust graffiti detector.

2.2. Step 1: Community Data Collection

Collecting and gathering suitable visual data was the first step in a successful implementation of the technique. Similar to many innovative structural assessment methods based on computer-vision-based technology, images provided useful information containing many viewpoints, at frequent time intervals, and for many locations. However, this application was different compared with typical damage-detection algorithms, as graffiti is an obvious and visible object in an image. Hence, while individual images and their qualities are considered to be more critical in damage-detection algorithms, overall image dataset variety and volume were more important for this application, including collection frequency and inclusiveness of the key structures.

Once the data were collected, our approach was to use the structure-from-motion (SfM) technique to establish a geometric relationship between images through orthophotos of façades of the TOC. The technical details of SfM are provided in the next subsection. In SfM, sufficient image overlap between images from various distances and orientations was required, although the specific amount of overlap needed varied depending on the image quality (e.g., resolution, noise, occlusions). Consequently, the collection of a large number of images was essential to this technique. Of course, such a large number of images having high temporal and spatial resolutions were necessary for monitoring the TOC.

Involving citizens can provide a major breakthrough in this type of data requirement. Crowdsourcing to obtain a sufficient quantity of visual data would satisfy this condition. A variety of ordinary individuals such as residents, workers, tourists, and visitors present in a community were encouraged to capture the necessary images. Smartphones are broadly used around the world for capturing such data, and the data can be uploaded to any readily available cloud service (e.g., Google Drive, Dropbox, Box). Social-media technologies (e.g., Instagram, Facebook, and YouTube) further accelerated this mechanism for broad data capture by making many resources of visual data available online [40]. Specific crowdsourcing strategies for visual assessment were discussed in our previous work, and are not repeated here [41,42]. For best results, the requirement in this step was the collection of a large number of images to cover the relevant surfaces of the TOC that may have been exposed to graffiti. A set of images was required to be collected within each assessment period to update the most recent scene of the TOC for detecting graffiti markings.

2.3. Step 2: Façade Image (Orthophoto) Generation

In this step, instead of searching for graffiti markings on individual images, generating a high-resolution façade image (orthophoto) of the TOC enabled the detection and localization of up-to-date graffiti presence. The full process would include estimating their sizes and locations or potentially implementing a remote monitoring system. The orthophoto was generated by projecting multiple images on a façade plane, while also removing perspective distortion [43–45]. It had a uniform scale in each direction defining its plane, so it truly represented an accurate view of the façades of the TOC. The wall planes of the TOC were automatically estimated by fitting a plane after a point cloud of the TOC was first constructed. Then, each image was projected onto that plane, followed by typical stitching and blending procedures. The number of orthophotos here depended on the number of façades in the TOC to be assessed. In Figure 4, we describe the step-by-step process needed to generate the necessary orthophotos from collected visual data that satisfied the data quantity and overlap requirements.

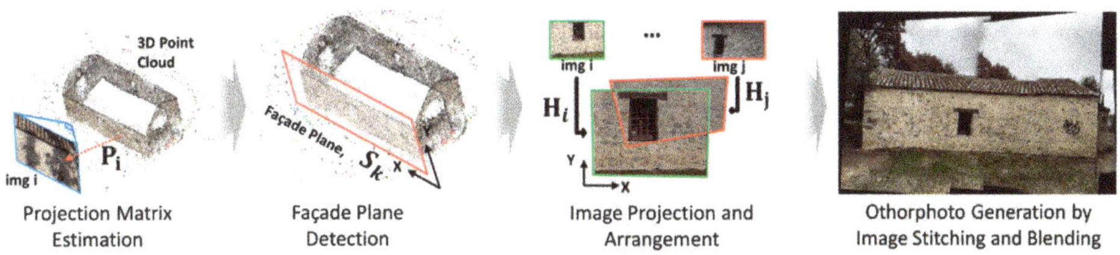

Figure 4. Several high-resolution façade images (orthophotos) generated for each TOC by automatically detecting major planes from a point cloud generated from collected images.

First, a projection matrix for each image was estimated. It was expressed in homogeneous coordinates, which was a 3 × 4 matrix for an ideal pinhole camera. The projection matrix was derived while considering both intrinsic parameters (focal length, principal points, and camera center) and extrinsic parameters (a relative camera location in 3D and rotation angle) of the image [25–27]. This process, which included large, iterative computations, could be greatly accelerated by SfM by computing shared keypoints between images [28]. The two important outputs generated from this step were: (i) a projection matrix of each image; and (ii) a point cloud describing the 3D scene of the TOC. By satisfying the data requirements outlined in Step 1, SfM yielded those two outputs without prior knowledge of either the geolocation or camera parameters. Considering each (i-th) image, the following relation held:

$$\mathbf{P}_i = \mathbf{K}_i[\ \mathbf{R}_i \quad \mathbf{T}_i\ ] \tag{1}$$

where $\mathbf{P}$ is the projection matrix (3 × 4 matrix), and the rotation matrix $\mathbf{R}$ (3 × 3 matrix) and translation matrix $\mathbf{T}$ (3 × 1 vector) represent the camera orientation and translation, respectively. $\mathbf{K}$ represents an upper triangular camera calibration matrix of the form:

$$K = \begin{bmatrix} \alpha_u & s & u_0 \\ 0 & \alpha_v & v_0 \\ 0 & 0 & 1 \end{bmatrix}$$

where α_u and α_v are scale factors, s is skew, and $\mathbf{u}_0 = [\ u_0 \quad v_0\ ]^{\mathbf{T}}$ is the principal point of the image. The relative locations and orientation of each image could be simply expressed with the projection matrix, $\mathbf{P}$, as expressed in Equation (1) [27].

Second, the façade planes on the TOC were estimated. The goal in this step was to automatically search for the major planes in the point cloud onto which the collected images were to be projected. The feature matches, in most cases, were the ones on the surface of the TOC. Thus, this can be allowed as a simple plane-fitting problem on a set of points in space. Here, we integrated the Maximum Likelihood Estimation Sample and Consensus (MLESAC) algorithm, which is widely used in estimating either a complex surface or a more general manifold from point data [46]. MLESAC follows the same sampling strategy as the Random Sample Consensus (RANSAC) algorithm, but more weight is placed on inlier samples by maximizing the likelihood for a better fitting. The façade plane estimated with the MLESAC was denoted as $\mathbf{S}$ (a 1×4 vector). Considering the estimated $\mathbf{S}$, the keypoint match lying on the plane $\mathbf{S}$ was denoted as $\mathbf{X}$. In this way, numerous orthophotos could be generated for every plane that existed in the TOC. Users could set the number of planes to be estimated; thus, orthophotos were generated only for a true façade in the TOC. Here, the number of façades to be estimated, j, could either be assigned or was set through an appropriate threshold in the process. Considering such factors, the equation of the plane in vector form is:

$$\mathbf{S}_j\mathbf{X}_j = \mathbf{0} \tag{2}$$

where the façade is expressed as $\mathbf{S} = \begin{bmatrix} a & b & c & 1 \end{bmatrix}$, which is the model for MLESAC. A major plane was estimated based on counting the number of $\mathbf{X}_j$ satisfying Equation (2). Here, the j number of major planes were finally chosen from the point cloud.

Third, to project each image on the major planes detected, a homography matrix, $\mathbf{H}$, was computed from the projection matrix, $\mathbf{P}$, and the estimated plane, $\mathbf{S}$. The homography matrix was a 3×3 matrix providing a planar projection transformation between two 2D images. This matrix thus provided a one-to-one mapping between the images and the estimated plane [27]. To do this, the homography relationship needed to transform between a *pixel* coordinate (frame) and *ortho* coordinate ($\mathbf{S}$) system (expressed in the x–y coordinate system). Computing a rotation coefficient, r, enabled homography from the projection matrix. This derivation was described in detail in [45]. Considering each image, i, and each estimated plane, j, this relationship can be represented as:

$$\mathbf{H}_{ij} = \begin{bmatrix} \mathbf{p}_i^1 & \mathbf{p}_i^2 & r_j\mathbf{p}_i^3 + \mathbf{p}_i^4 \end{bmatrix} \tag{3}$$

where $\mathbf{p}_i^k$ is the kth column of $\mathbf{p}_i$.

Finally, the set of images was projected onto the estimated planes using the homography matrices, which formed the orthophoto. In Equation (3), each image has j homography matrices, as we have j estimated planes. Thus, we needed to determine the proper façade plane (or, the homography matrix) onto which each image had to be projected. This process may result in regions, especially in oblique images, that extend beyond a given façade (which are not in the plane) and would give the appearance of having large distortions when the perspective distortion is removed with respect to the façade plane. To avoid such cases, we used only a subset of the images that included those that were captured relatively parallel and close to the façade plane. We denoted this subset of images used as the *orthoSet*. A threshold value was set to automatically identify these images. The images in the *orthoSet* were chosen based on the angle, defined here as Θ, as well as a distance, defined as δ, between each image plane and the estimated plane. Based on the experimental verification we performed in Section 3, we suggested that these thresholds be set such that that Θ be smaller than $15°$ and δ be less than 5 m. From this point on, we followed the procedure used by Allène et al. to blend the images and generate an orthophoto using gain compensation and multiband blending [29].

2.4. Step 3: Graffiti Detection on Orthophotos

As a final step, graffiti detection took place on each orthophoto generated in Step 2. The use of an orthophoto in detecting graffiti had several advantages. The TOC could be completely captured with several orthophotos, thus encapsulating any overall graffiti presence including their locations and sizes. This would not have been possible only using individual crowd-sourced images due to their disconnection or repetition of a particular scene, especially for large TOCs. The efficacy of this approach increased when considering long-term maintenance from a lifecycle assessment perspective, as depicted in Figure 5. With public online data resources providing abundant and up-to-date visual data, we could easily track graffiti presence by simply comparing a few orthophotos over time. Visual data may be the simplest and most cost-efficient means to document visual information related to the condition of properties in the community, especially historical structures that are popular attractions.

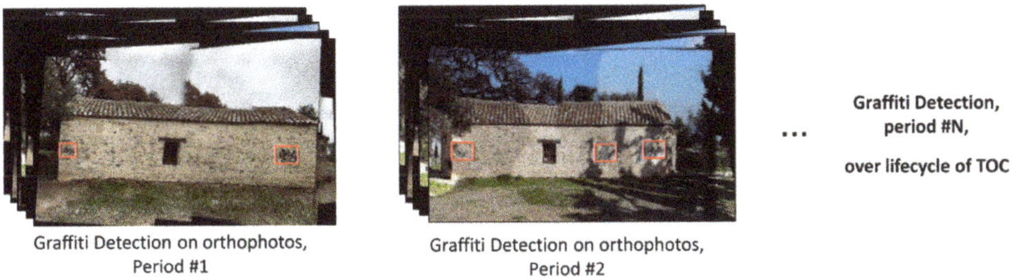

Figure 5. The robust graffiti detector was applied to each of the orthophotos, providing an overall presence of graffiti, including the location and size of each mark.

3. Experimental Verification

We demonstrated the graffiti detection technique by applying it to a real-world structure exhibiting various types of graffiti. A target structure, the Church of Agios Nikolaos, Kantza, Greece, is a cultural heritage building. Since we noticed that its exterior condition is one of the major interests for maintenance, we regarded the Church of Agios Nikolaos as the TOC for verification of the proposed method through this case study. As a preprocessing step, a graffiti detector was trained using a graffiti data set established within the STORM project. Then, the technique was demonstrated on our TOC, where several graffiti markings had already been placed. For this implementation, visual data were collected twice within an eight-month interval to highlight the capability of structure monitoring.

3.1. *Graffiti Detector*

The STORM project developed a graffiti database for various purposes, including the monitoring of cultural heritage sites in European countries [32]. The database includes a wide variety of scenes containing graffiti, such as markings drawn on walls, benches, and statues from Athens, Greece, and its suburbs. A total of 1022 images that were collected from various locations and viewpoints were used. Multiple graffiti markings were labeled from each image for augmenting the ground-truth dataset for the graffiti detector in this case study.

Our approach to training the graffiti detector was to implement Faster R-CNNs [31]. Graffiti markings can overlap, and their boundaries may be visually ambiguous. Thus, having an accurate and uniform labeling criteria played a crucial role. We provided guidelines to obtain consistency in creating the ground-truth dataset. First, graffiti markings should be completely surrounded by a rectangular bounding box. Second, the bounding box should be drawn for each graffiti (within recognizable range), and there can be more than one graffiti marking captured on an image. Multiple graffiti markings were separated within recognizable range considering shape, color, and drawing style, along with allowing overlap between the bounding boxes. By following these two guidelines, a total of 1682 graffiti markings from the 1022 images were labeled by three different individuals to build a ground-truth dataset for the graffiti detector. Cross-validation was conducted to generate this training data set to input into the Faster R-CNN algorithm. Sample images from the database are shown in Figure 6.

Figure 6. Graffiti dataset provided by the STORM project and used for training the graffiti detector: (**a**) sample images from the database of graffiti scenes in a community; and (**b**) samples of graffiti markings in our ground-truth dataset identified by three individuals on our team with cross-validation.

In training the graffiti detector, we used a residual network model, called ResNet-101, implemented in the Faster R-CNN library [31]. The Faster R-CNN and its region proposal network (RPN) significantly reduced runtime by implementing object feature map computation on a GPU, providing an effective engineering solution. We deployed and exploited the Tensorflow library and implemented the Faster R-CNN algorithm in Python script [47]. ConvNets and fully connected layers were initialized by zero-mean Gaussian with a standard deviation of 0.01 and 0.001, respectively. The hyperparameters were the same used in [48]. Momentum and weight decay for training the R-CNN and RPN networks were set to 0.9 and 0.0005, respectively. The learning rate, defined as the amount the weights were adjusted with respect to the loss gradient, was set to 0.001. To achieve faster convergence of the gradient descent, we set the momentum to 0.9. The decay weight was also controlled to avoid overfitting by changing the weight in proportion to each patch size.

In the model training, we split the dataset 80%/20% into training and testing, respectively: 818 images for training and 204 images for testing. Then, we set aside the test data set, and randomly choose 10% of the training dataset to be the validation data set for cross-validation of the model. Since multiple graffiti markings existed in a single image, the total number of the 1200 graffiti markings were considered from the 204 test images. A PC workstation with a Xenon E5-2620 CPU clocked at 3.4 GHz, 12 GB of RAM, and a 64-bit operating system were used for this process. Two GPUs, an NVIDIA Titan X and an NVIDIA Tesla k40 with a total of 24 GB of VRAM, were used in the training. The runtime for the process varied depending on the size of graffiti data set, as well as computational resources available. For our graffiti detector, with the settings listed above and 80% of the total data of 818 images, the total training runtime averaged 135 min. Figure 7 shows example results with test images on which multiple complicated graffiti markings existed within the scene image. As shown in the confusion matrix in Table 1, we finally achieved 88% accuracy of the model with 204 images in the test.

Figure 7. Example results with test images on which multiple complicated graffiti markings existed within the scene image.

Table 1. Confusion matrix for calculating the accuracy of the trained R-CNN.

		Predicted Class	
		Positive	**Negative**
Actual Class	Positive	482	83
	Negative	61	574

3.2. Description of the Target Object in a Community (TOC): Historical Structure in Kantza, Greece

The TOC in this case study, the Church of Agios Nikolaos (Figure 8a) is a cultural heritage building and historical structure, and its exterior condition is of concern for city maintenance. This 400-year-old building is 5 m × 12 m and is currently in use in Kantza-Pallini, Greece. It was originally built in 1592 AD, and was restored in 1872. As a regional landmark structure, together with a colossal lion sculpture made of Penteli mountain marble located at the NW of the temple, they offer various crowdsourcing opportunities that could involve both citizens and online resources. As shown in Figure 8b, black and green-colored graffiti markings were present on all four of its walls, where its light-colored surfaces made the graffiti even more obtrusive and ugly.

(a) (b)

Figure 8. (**a**) The Church of Agios Nikolaos in Kantza, Greece, the historical structure used in this case study; and (**b**) samples of existing graffiti markings present.

Exploring appropriate and stable crowdsourcing resources was important in this method. In this work, we focused on the technical implementation procedures and assumed proper crowdsourcing resources were already retained. Here, we assumed that a large volume of visual data was collected as a part of a series of tasks in a scavenger-hunt-like game application that users were asked to complete [49], and there were a sufficient quantity and variability to satisfy the minimum image overlap criterion in order to perform the SfM procedure described in Section 2 [43]. The game application was developed in the scope of the STORM project [50]. Thus, the community data collection step here was done by our team members, not random individuals. Furthermore, to demonstrate the potential for enabling a lifecycle assessment perspective, we collected data twice within an eight-month interval, including 268 images in May 2018 and 337 images in January 2019.

3.3. Façade Image (Orthophoto) Generation

To estimate the projection matrix for each of the collected images, we used VisualSfM [51]. VisualSfM is an open-source SfM software that provides a user-friendly graphical user interface (GUI) to monitor the intermediate steps of the SfM process, such as keypoint feature matching and camera pose estimation. VisualSfM highly improved the speed of the SfM computations by implementing the SiftGPU Library and parallel processing using graphics processing units (GPUs) [44–51]. Finally, VisualSfM yielded two outputs that were important in this work; namely, the projection matrix of each collected image and the point cloud of the Church of Agios Nikolaos. The MLESAC algorithm for façade plane estimation from the point cloud and orthophoto generation was implemented in MATLAB [52,53]. The average processing time for generating a total of four complete orthophotos from the 182 images was 1.5 h using the PC workstation mentioned earlier.

Since the Church of Agios Nikolaos has four major obvious façades, we did not need to manually assign the parameter j. If a TOC has a complicated shape of façades, the value of j can be set manually. The algorithm here was able to successfully estimate the four façade planes, and then the four orthophotos were generated. Figure 9 depicts the point cloud and one orthophoto generated in May 2018. We set $\Theta < 15°$ and $\delta < 5$ m to reduce the number of projection images for each orthophoto. These threshold values are subject to different cases, but for applications involving buildings in a community, they are likely a good choice. A total of 98 images were automatically chosen for reconstructing all four façades. A representative result showing an orthophoto of a façade plane is provided in Figure 9b. Some distortions and discontinued scenes were observed at the edges, but these were outside of façade area, roof, and ground. Thus, they did not disrupt the graffiti detection process because those distortions were outside of the façade.

Figure 9. Steps in the orthophoto generation process for the Church of Agios Nikolaos: (**a**) the point cloud generated and its four major façade planes; and (**b**) a representative orthophoto generated from $\mathbf{S}_1$.

3.4. Graffiti Detection on Orthophotos

After generating orthophotos from all four façade planes of the TOC, the graffiti detector we trained in Section 3.1 was then applied to each orthophoto. The results of the graffiti-detection process on each façade plane on the Church of Agios Nikolaos are shown in Figure 10. The detected graffiti are marked with green rectangular bounding boxes. All graffiti presented on the façades were successfully identified. Since the individual images were collected at various locations and viewpoints, causing significant brightness or exposure discrepancies, the lighting differences over the region of the orthophoto were not fully compensated. In addition, the overall quality of the orthophotos was inconsistent because they were highly dependent on the quality of the input images. Nevertheless, graffiti markings were still observed and identified by the graffiti detector. We conducted data collection in two different time windows. We noticed that all graffiti markings were successfully detected on each of orthophotos, and readily knew that there was no addition or removal. This demonstrated that the trained detector was robust enough, and was insensitive to the environmental variation in the background scene. The graffiti distribution was shown in a single view, which will also assist in assessing the overall exterior condition and directing maintenance teams to the proper location. The processing for this step took an average of 1.4 min for each of the four orthophotos.

Figure 10. Graffiti-detection results from the orthophotos generated for the four façades of the Church of Agios Nikolaos based on data collected in (**a**) May 2018 and (**b**) January 2019.

4. Conclusions

This work successfully demonstrated an automated graffiti-detection technique for cultural heritage buildings. Data from public sources was combined and used to detect, localize, and quantify graffiti on key buildings in a community. The required visual data can be collected by ordinary citizens periodically to enable more efficient lifecycle assessment of historical structures. The output of the method was the location and size of each graffiti marking, indicated by a bounding box provided on the orthophoto. As a part of this process, a robust graffiti detector was also trained here using images of graffiti collected from real-world structures, and the training data were shared publicly for other researchers to access [32]. The feasibility of the overall method was demonstrated with a historical structure in the European community: the Church of Agios Nikolaos, located in Kantza, Greece. Images were collected in two different time windows and in weather conditions to demonstrate the capability of the technique for the use of lifecycle assessment.

To the authors' best knowledge, no vision-based graffiti detection and quantification technique exists in the literature or industry. A completely automated technique was fully demonstrated here through a real-world case study. Some practical challenges do still exist in the technique that should be mentioned. The method is not applicable to cases in which the entire façade is contaminated with graffiti, or cases in which a significant curvature exists in the outer walls of the building being assessed, which creates distorted orthophoto generation.

5. Future Work

There are improvements to be made during the orthophoto generation in order to make it more robust and add the capability to assess structures with more irregular shapes. The step of façade-plane estimation benefited from the already rectangular and smooth surfaces of the structure used during experimentation. If the structure had a significant curvature (such as a toroidal shape), restraining the number of surfaces to be estimated and identifying planes tangent to the 3D point cloud would be required.

Author Contributions: Conceptualization, J.C.; Methodology, J.C., L.T., and C.M.Y.; Software, R.O., J.C. and C.M.Y.; Validation, N.-J.J., S.J.D. and C.M.Y.; Formal Analysis, R.O., S.J.D. and C.M.Y.; Investigation, J.C.; Resources, L.T., P.K., and P.C.; Data Curation, R.O., J.C., A.L., and X.L.; Writing—Original Draft Preparation, J.C.; Writing—Review and Editing, S.J.D. and P.C.; Visualization, J.C.; Supervision, S.J.D. and P.C. All authors have read and agreed to the published version of the manuscript.

Funding: This work supported by the Civil, Mechanical and Manufacturing Innovation (CMMI) Program of the National Science Foundation (Grant No. NSF 1645047), the European Union's Horizon 2020 research program under the STORM project (Grant No. 700191), and the National Research Foundation of Korea funded by the Korean government (MSIT) (Grant No. NRF-2021R1G1A1012298).

Institutional Review Board Statement: Not applicable.

Informed Consent Statement: Not applicable.

Acknowledgments: We would like to thank the Ephorate of Antiquities of Eastern Attica for giving us permission to use the photos of the Church of Agios Nikolaos in order to complete our study and use case, and the NVIDIA Corporation for the donation of a high-end GPU board. Finally, the authors would like to extend their appreciation to Alexandros Tzitamidis for annotating the graffiti-detection data set.

Conflicts of Interest: The authors declare no conflict of interest.

References

1. Home. Oxford English Dictionary. Available online: https://www.oed.com/ (accessed on 25 February 2022).
2. Holocaust Monument in Central Athens Vandalized with Graffiti. Naftemporiki.gr. 2017. Available online: https://www.naftemporiki.gr/story/1300966/holocaust-monument-in-central-athens-vandalized-with-graffiti (accessed on 25 February 2022).
3. Armstrong, J.S. The Graffiti Problem. University Library of Munich, Germany, 0412035. 2004. Available online: https://ideas.repec.org/p/wpa/wuwpgt/0412035.html (accessed on 25 February 2022).

4. Sanmartín, P.; Cappitelli, F.; Mitchell, R. Current methods of graffiti removal: A review. *Const. Build. Mater.* **2014**, *30*, 363–374. [CrossRef]
5. White, R. Graffiti, Crime Prevention & Cultural Space. *Curr. Issues Crim. Justice* **2001**, *12*, 253–268. [CrossRef]
6. Halsey, M.; Young, A. The Meanings of Graffiti and Municipal Administration. *Aust. New Zeeland J. Criminol.* **2002**, *35*, 165–186. [CrossRef]
7. Macdonald, N. *The Graffiti Subculture: Youth, Masculinity, and Identity in London and New York*; Palgrave: New York, NY, USA, 2001.
8. Ferrell, J. *Crimes of Style: Urban Graffiti and the Politics of Criminality*; Garland: New York, NY, USA, 1993.
9. Ley, D.; Cybriwsky, R. Urban Graffiti as Territorial Markers. *Ann. Assoc. Am. Geogr.* **1974**, *64*, 491–505. [CrossRef]
10. Spencer, B.F.; Hoskere, V.; Narazaki, Y. Advances in Computer Vision-Based Civil Infrastructure Inspection and Monitoring. *Engineering* **2019**, *5*, 199–222. [CrossRef]
11. Jahanshahi, M.R.; Masri, S.F.; Padgett, C.W.; Sukhatme, G.S. An innovative methodology for detection and quantification of cracks through incorporation of depth perception. *Mach. Vis. Appl.* **2013**, *24*, 227–241. [CrossRef]
12. Ham, Y.; Han, K.K.; Lin, J.J.; Golparvar-Fard, M. Visual monitoring of civil infrastructure systems via camera-equipped Unmanned Aerial Vehicles (UAVs): A review of related works. *Vis. Eng.* **2016**, *4*, 118–261. [CrossRef]
13. Liu, X.; Dyke, S.J.; Yeum, C.M.; Bilionis, I.; Lenjani, A.; Choi, J. Automated Indoor Image Localization to Support a Post-Event Building Assessment. *Sensors* **2020**, *20*, 1610. [CrossRef]
14. Narazaki, Y.; Hoskere, V.; Eick, B.A.; Smith, M.D.; Spencer, B.F. Vision-based dense displacement and strain estimation of miter gates with the performance evaluation using physics-based graphics models. *Smart Struct. Syst.* **2019**, *24*, 709–721. [CrossRef]
15. Kong, X.; Li, J. Vision-Based Fatigue Crack Detection of Steel Structures Using Video Feature Tracking. *Comput. Aided Civ. Infrastruct. Eng.* **2018**, *33*, 783–799. [CrossRef]
16. Gong, J.; Azambuja, M. Visualizing Construction Supply Chains with Google Cloud Computing Tools. In Proceedings of the International Conference on Sustainable Design, Engineering, and Construction, ICSDEC, Fort Worth, TX, USA, 7–9 November 2012; pp. 671–678. [CrossRef]
17. Choi, J.; Park, J.A.; Dyke, S.J.; Yeum, C.M.; Liu, X.; Lenjani, A.; Bilionis, I. Autonomous image localization for visual inspection of civil infrastructure. *Smart Mater. Struct.* **2017**, *26*, 035051. [CrossRef]
18. JChoi, J.; Park, J.A.; Dyke, S.J.; Yeum, C.M.; Liu, X.; Lenjani, A.; Bilionis, I. Similarity learning to enable building searches in post-event image data. *Comput. Aided Civ. Inf.* **2022**, *37*, 261–275. [CrossRef]
19. Farrar, C.R.; Worden, K. *Structural Health Monitoring: A Machine Learning Perspective*; John Wiley & Sons: Hoboken, NJ, USA, 2012.
20. Yeum, C.M.; Choi, J.; Dyke, S.J. Automated region-of-interest localization and classification for vision-based visual assessment of civil infrastructure. *Struct. Health Monit.* **2019**, *18*, 675–689. [CrossRef]
21. Lenjani, A.; Dyke, S.J.; Bilionis, I.; Yeum, C.M.; Kamiya, K.; Choi, J.; Liu, X.; Chowdhury, A.G. Towards fully automated post-event data collection and analysis: Pre-event and post-event information fusion. *Eng. Struct.* **2020**, *208*, 109884. [CrossRef]
22. Wu, R.; Singla, A.; Jahanshahi, M.R.; Bertino, E.; Ko, B.J.; Verma, D. Pruning deep convolutional neural networks for efficient edge computing in condition assessment of infrastructures. *Comput. Aided Civ. Infrastruct. Eng.* **2019**, *34*, 774–789. [CrossRef]
23. Koch, C.; Paal, S.; Rashidi, A.; Zhu, Z.; König, M.; Brilakis, I. Achievements and challenges in machine vision-based inspection of large concrete structures. *Adv. Struct. Eng.* **2014**, *17*, 303–318. [CrossRef]
24. Mondal, T.G.; Jahanshahi, M.R.; Wu, R.; Wu, Z.Y. Deep learning-based multi-class damage detection for autonomous post-disaster reconnaissance. *Struct. Control Health Monit.* **2020**, *27*, e2507. [CrossRef]
25. Kraus, K. *Photogrammetry: Geometry from Images and Laser Scans*; Walter de Gruyter: Berlin/Heidelberg, Germany, 2011.
26. Snavely, N.; Seitz, S.M.; Szeliski, R. Modeling the World from Internet Photo Collections. *Int. J. Comput. Vis.* **2008**, *80*, 189–210. [CrossRef]
27. Hartley, R.; Zisserman, A. *Multiple View Geometry in Computer Vision*; Cambridge University Press: Cambridge, UK, 2003.
28. Westoby, M.J.; Brasington, J.; Glasser, N.F.; Hambrey, M.J.; Reynolds, J.M. Structure-from-Motion' photogrammetry: A low-cost, effective tool for geoscience applications. *Geomorphology* **2012**, *179*, 300–314. [CrossRef]
29. Allene, C.; Pons, J.-P.; Keriven, R. Seamless Image-Based Texture Atlases Using Multi-Band Blending. In Proceedings of the 2008 19th International Conference on Pattern Recognition, Tampa, FL, USA, 8–11 December 2008; pp. 1–4. [CrossRef]
30. BYZANTINA MNHMEIA ATTIKHΣ. Available online: http://byzantineattica.eie.gr/byzantineattica/ (accessed on 25 February 2022).
31. Ren, S.; He, K.; Girshick, R.; Sun, J. Faster R-CNN: Towards Real-Time Object Detection with Region Proposal Networks. In *Advances in Neural Information Processing Systems 28*; Cortes, C., Lawrence, N.D., Lee, D.D., Sugiyama, M., Garnett, R., Eds.; Curran Associates, Inc.: La Jolla, CA, USA, 2015; pp. 91–99. Available online: http://papers.nips.cc/paper/5638-faster-r-cnn-towards-real-time-object-detection-with-region-proposal-networks.pdf (accessed on 25 February 2022).
32. Patrikakis, C.; Kasnesis, P.; Toumanidis, L.; Tzitamidis, A. zenodo.org; *STORM Graffiti/Tagging Detection Dataset*; CERN: Meyrin, Switzerland, 2019. [CrossRef]
33. Uijlings, J.R.R.; van de Sande, K.E.A.; Gevers, T.; Smeulders, A.W.M. Selective Search for Object Recognition. *Int. J. Comput. Vis.* **2013**, *104*, 154–171. [CrossRef]
34. Girshick, R.; Donahue, J.; Darrell, T.; Malik, J. Region-Based Convolutional Networks for Accurate Object Detection and Segmentation. *IEEE Trans. Pattern Anal. Mach. Intell.* **2016**, *38*, 142–158. [CrossRef] [PubMed]

35. Girshick, R. Fast R-CNN. 2015. Available online: http://openaccess.thecvf.com/content_iccv_2015/html/Girshick_Fast_R-CNN_ICCV_2015_paper.html (accessed on 25 February 2022).
36. Gkioxari, G.; Johnson, J.; Malik, J. Mesh r-cnn. In Proceedings of the IEEE/CVF International Conference on Computer Vision, Seoul, Korea, 27 October–2 November 2019; pp. 9785–9795.
37. Hinton, G.E.; Salakhutdinov, R.R. Replicated Softmax: An Undirected Topic Model. In *Advances in Neural Information Processing Systems 22*; Bengio, Y., Schuurmans, D., Lafferty, J.D., Williams, C.K.I., Culotta, A., Eds.; Curran Associates, Inc.: Nice, France, 2009; pp. 1607–1614. Available online: http://papers.nips.cc/paper/3856-replicated-softmax-an-undirected-topic-model.pdf (accessed on 25 February 2022).
38. Liu, W.; Anguelov, D.; Erhan, D.; Szegedy, C.; Reed, S.; Fu, C.Y.; Berg, A.C. SSD: Single Shot MultiBox Detector. In Proceedings of the Computer Vision—ECCV 2016, Cham, Switzerland, 11–14 October 2016; pp. 21–37. [CrossRef]
39. Redmon, J.; Farhadi, A. YOLO9000: Better, Faster, Stronger. 2017. Available online: http://openaccess.thecvf.com/content_cvpr_2017/html/Redmon_YOLO9000_Better_Faster_CVPR_2017_paper.html (accessed on 25 February 2022).
40. Saxton, G.D.; Oh, O.; Kishore, R. Rules of Crowdsourcing: Models, Issues, and Systems of Control. *Inf. Syst. Manag.* **2013**, *30*, 2–20. [CrossRef]
41. Follett, R.; Strezov, V. An Analysis of Citizen Science Based Research: Usage and Publication Patterns. *PLoS ONE* **2015**, *10*, e0143687. [CrossRef]
42. Choi, J.; Dyke, S.J. CrowdLIM: Crowdsourcing to enable lifecycle infrastructure management. *Comput. Ind.* **2020**, *115*, 103185. [CrossRef]
43. Agisoft Metashape. Available online: https://www.agisoft.com/ (accessed on 25 February 2022).
44. Wu, C. Critical Configurations for Radial Distortion Self-Calibration. 2014, p. 25. Available online: http://openaccess.thecvf.com/content_cvpr_2014/html/Wu_Critical_Configurations_For_2014_CVPR_paper.html (accessed on 25 February 2022).
45. Choi, J.; Yeum, C.M.; Dyke, S.J.; Jahanshahi, M.R. Computer-Aided Approach for Rapid Post-Event Visual Evaluation of a Building Façade. *Sensors* **2018**, *18*, 3017. [CrossRef] [PubMed]
46. Torr, P.H.S.; Zisserman, A. MLESAC: A New Robust Estimator with Application to Estimating Image Geometry. *Comput. Vis. Image Underst.* **2000**, *78*, 138–156. [CrossRef]
47. TensorFlow. Available online: https://www.tensorflow.org/?hl=ko (accessed on 25 February 2022).
48. Chen, X.; Gupta, A. An Implementation of Faster RCNN with Study for Region Sampling. *arXiv* **2017**, arXiv:170202138 Cs. Available online: http://arxiv.org/abs/1702.02138 (accessed on 25 February 2022).
49. Toumanidis, L.; Bocaj, E.; Kasnesis, P.; Patrikakis, C.Z. Supporting Cultural Heritage Preservation Through Game-Based Crowdsourcing. In *Strategic Innovative Marketing and Tourism*; Springer: Cham, Switzerland, 2019; pp. 989–997. [CrossRef]
50. Kasnesis, P.; Kogias, D.G.; Toumanidis, L.; Xevgenis, M.G.; Patrikakis, C.Z.; Giunta, G.; Calsi, G.L. An IoE Architecture for the Preservation of the Cultural Heritage: The STORM Use Case. Harnessing the Internet of Everything (IoE) for Accelerated Innovation Opportunities. 2019. Available online: www.igi-global.com/chapter/an-ioe-architecture-for-the-preservation-of-the-cultural-heritage/221288 (accessed on 25 February 2022).
51. VisualSFM: A Visual Structure from Motion System. Available online: http://ccwu.me/vsfm/ (accessed on 25 February 2022).
52. MATLAB—MathWorks—MATLAB & Simulink. Available online: https://www.mathworks.com/products/matlab.html (accessed on 25 February 2022).
53. Kuzmin, Y.P.; Korytnik, S.A.; Long, O. Polygon-based true orthophoto generation. *Int. Arch. Photogramm. Remote Sens. Spat. Inf. Sci.* **2004**, *35*, 529–531.

applied sciences

MDPI

Article

Implementation of AHP Methodology for the Evaluation and Selection Process of a Reverse Engineering Scanning System

Juraj Beniak *, Ľubomír Šooš, Peter Križan and Miloš Matúš

Faculty of Mechanical Engineering, Slovak University of Technology in Bratislava, Nam. Slobody 17, 812 31 Bratislava, Slovakia; lubomir.soos@stuba.sk (Ľ.Š.); peter.krizan@stuba.sk (P.K.); milos.matus@stuba.sk (M.M.)

* Correspondence: juraj.beniak@stuba.sk

Abstract: Generally speaking, the proper selection of a suitable system for various uses is key to its full use in practice. In all areas, there is a large number of technologies, equipment, and systems to choose from, so it is necessary to determine the individual parameters and their weight, which are important for selection. In the field of reverse engineering, several technological devices are particularly expensive, and the selection of one will influence the long-term functioning of the system. Reverse engineering systems are widely used for the registration and documentation of historical objects in the sense of cultural heritage, and the presented scanning systems are suitable for this purpose. In this case, the selection of a scanning system is discussed. This paper deals with the methodology of selecting the most suitable reverse engineering system by the method of pairwise comparison of expert evaluation criteria (analytical hierarchical process (AHP)). This paper contains a comparison of several systems and the selection of the most suitable solution for the particular company.

Keywords: reverse engineering; evaluation process; 3D scanners

Citation: Beniak, J.; Šooš, Ľ.; Križan, P.; Matúš, M. Implementation of AHP Methodology for the Evaluation and Selection Process of a Reverse Engineering Scanning System. *Appl. Sci.* **2021**, *11*, 12050. https://doi.org/10.3390/app112412050

Academic Editors: Marek Milosz and Jacek Kęsik

Received: 14 November 2021
Accepted: 14 December 2021
Published: 17 December 2021

Publisher's Note: MDPI stays neutral with regard to jurisdictional claims in published maps and institutional affiliations.

1. Introduction

The aim of the research was to design a non-contact method for obtaining a digital 3D model of a produced part, so as to measure the quality of production. This is an operation that has only been carried out using contact devices so far. These excel in their greater accuracy, but their measuring speed and ability to measure free-form products are low or insufficient [1,2]. For this purpose, it was proposed to use optical systems for scanning, and then to compare the data obtained from actual manufactured products with those data from 3D digital models that were used for production [3,4].

Another area of application of reverse engineering systems, such as scanning systems, is the documentation of historical objects and monuments that are part of the cultural heritage of human society [5].

As such, optical systems have come a long way during their development, and their accuracy, in which they have been lagging behind, is beginning to approach that of a coordinate measuring instrument [6–8]. The scope of their use is wide, mainly for operations with a large variety of products. Optical systems can shorten the time of measurement, as well as the evaluation of measured data, which significantly streamlines the work of the employees of the relevant department. A convincing argument for the implementation of scanning devices is also the cost saving, simplification and acceleration of the quality control process. Above all, there is an overall increase in the level of quality [9–11].

The development of technologies in reverse engineering opens new horizons in the areas of quality assurance and control. Great progress is being made with non-contact devices, the accuracy of which is beginning to approach that of conventional contact coordinate measuring instruments. One of their main advantages is the scanning speed; the systems can scan the measured object several times faster, which ultimately means increased productivity and the ability to control a larger number of products. Another advantage of contactless scanners is the easier and faster processing, when comparing the measured data with a CAD model, which, again, brings an increase in the productivity of quality personnel, scope to reduce labor costs, and extends quality control functions to larger production volumes [12–14]. The basic concept of the reverse engineering process is shown in Figure 1.

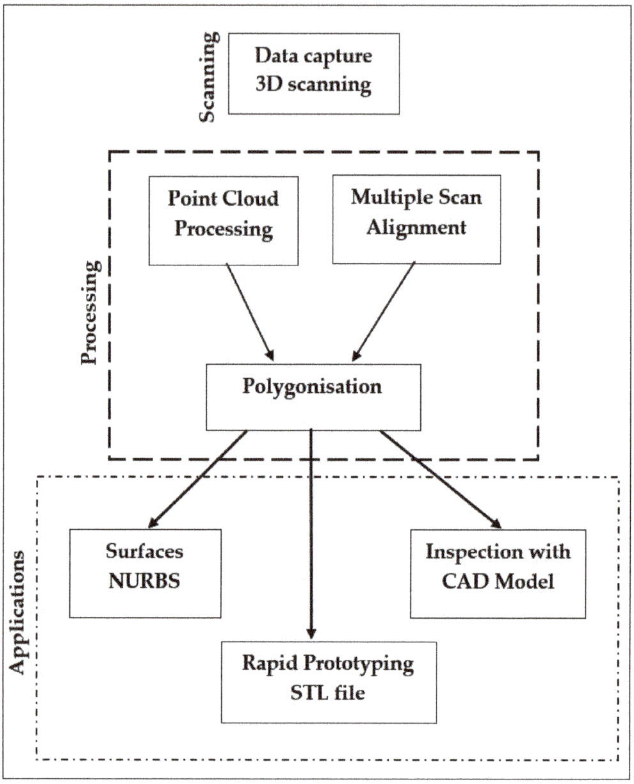

Figure 1. Reverse engineering process—basic concept.

The procurement of such technology for scanning real objects is demanding, in terms of finances and the number of alternatives available on the market. Any decision of one expert is only subjective, and may not take into account all the aspects necessary for a proper evaluation of the characteristics of the available equipment and the selection of the most appropriate solution [15]. Therefore, it is appropriate to use scientific methods to assess the selected factors, preferably with the help of experts from several fields, who have a different perspective on evaluation [16].

This article deals with the system of selecting a suitable scanning device for reverse engineering using the pairwise comparison method and AHP. Pairwise comparison is a scientific method of analyzing a number of different objects or subjects to determine if they are significantly different from one another.

There are several different methods, which are based on the same principle—assessing several variants of solving a given problem according to the selected criteria, and determining the order of these variants. The individual methods differ according to how the so-called weight of the individual criteria and the degree to which the individual variants of the solution meet the selected criteria are numerically evaluated [17,18].

DMM (decision matrix method) is considered to be a basic method (it can have several solution variants) [19–21]. One of the variants consists of evaluating the weight (importance) of individual criteria with a point scale from 1 to 10, so that level 1 is assigned to the smallest weight and level 10 to the largest weight. The same scale also evaluates how individual variants of the solution meet the selected criteria, i.e., grade "1"—does not match up, to "10"—matches perfectly.

FDMM (modified decision matrix method) partially eliminates the disadvantages of DMM. The weights of individual criteria, as well as the evaluation of the ability of the variants to meet individual criteria, are determined by so-called pairwise comparison [22–24]. This means that when comparing the two criteria, the more important (more important for decision making) criterion is rated "1", and the less important criterion is "0". Similarly, when evaluating how two variants meet the selected evaluation criteria, the more satisfactory variant is rated "1", and the variant rated worse is "0".

AHP (analytic hierarchy process) eliminates the shortcomings of DMM and FDMM to some extent. It is also based on a pairwise comparison of the degree of significance of the individual criteria and the degree to which the evaluated solution variants meet these criteria. However, the rating scale is much more complex [25,26]. The evaluation is based, in both cases (comparison of criteria and variants), on so-called "expert estimation", in which experts in the field compare the mutual influences of two factors. These are evaluated on the basis of the scale equal—weak—medium—strong—very strong, while the values (1–3–5–7–9) correspond to this verbal evaluation.

2. Materials and Methods

For the correct selection of a suitable hardware solution for reverse engineering scanning, it is necessary to perform a thorough analysis on the basis of which the selection will be made. From the available products, a group of devices was selected for this purpose, which will be subjected to comparison. These are the following devices:

- MetraSCAN 70 (a_1);
- MetraSCAN 70-R (a_2);
- HandyPROBE (a_3);
- Nikon XC65Dx(a_4);
- Nikon LC60Dx (a_5);
- Nikon LC15Dx (a_6);
- Metronor DUO (a_7);
- ATOS Triple Scan (a_8).

The MetraSCAN 70, MetraSCAN 70-R, and HandyPROBE (Figure 2) are manufactured by Creaform, a Canadian company. The MetraSCAN 70 and 70-R are almost identical optical scanners. The main difference is in the compatibility of the R version with the mechanical arm, but only with the KR5 Arc product line from KUKA. In this configuration, the measuring system is not affected by the accuracy of the arm. The second difference is their weight; the lighter version of MetraSCAN 70 is designed for manual handling, so it is characterized by high portability. A necessary part of both scanners is the C-Track 380 position evaluation system and the VXelements software (data collection), which can be connected to the Metrolog X4 i-Robot software with more sophisticated measurement evaluation tools.

Figure 2. Alternatives from Creaform.

The HandyPROBE product is a wireless CMM (coordinate measuring machine) system consisting of a scanning probe with built-in reference points. These are continuously and dynamically scanned by the C-Track 380 optical system, from which points are then obtained for position evaluation. The basic parameters of three of the scanning devices listed above are displayed in Table 1.

Table 1. Technical specification of Creaform products.

Parameter	MetraSCAN 70	MetraSCAN 70R	HandyPROBE
Price (EUR)	67,350	130,000	38,500
Weight (kg)	1.85 kg	1.85 kg	0.45 kg
Dimensions (mm)	282 × 250 × 282		204 × 159 × 97
Measuring speed (pts/s)	36,000		30
Volumetric accuracy (mm)	0.075		0.022
Spacing distance (mm)	152		N/A
Depth/width FOV (Field of View) (mm)	50/2 × 70		N/A
Operating humidity range (%)	10–90%		10–90%
Working temperature (°C)	15–40		15–40 °C

The following three laser scanners from Nikon can be combined with coordinate measuring systems: Nikon XC65Dx, Nikon LC60Dx and Nikon LC15Dx (Table 2). They are characterized by a relatively high accuracy, which is slightly affected by the accuracy of the CMM. The following additional hardware is required for the scanner application: the Renishaw PH10M (Q) motorized three-axis head and the optical probe driver, let us say, scanner. To generate movements of the measuring probe around the measured part, the Focus Scan software, which is an integral part of the Focus Inspection software, uses its CAD model. This software is also part of the investment package.

The accuracy of the Metronor DUO from the Norwegian company Metronor AS is comparable to that of the HandyPROBE product. The advantage of this wireless CMM system is its greater distance from the measured objects, allowing measurements to be performed on multiple workbenches without the need to move the optical scanning system. The product comes with PowerInspect software. The ATOS Triple Scan optical scanner uses a method of illuminating an object with structured light, specifically narrow-band blue light, which allows measurement independent of ambient light conditions. It uses two eight-megapixel cameras to detect changes in the shape of the raster, from which the shape and dimensions of the scanned part are then determined using factory software. Its slight disadvantage is the need to place reference points around the measured object. Basic technical information about the Metronor DUO and ATOS III Triple Scan devices can be found in Table 3.

Table 2. Technical specification of Nikon products.

Parameter	XC65Dx	LC60Dx	LC15Dx
Price (EUR)	97,130	65,110	64,850
Weight (kg)	0.44	0.39	0.37
Dimensions (mm)	155 × 86 × 142	N/A	100 × 104 × 58
Measuring speed (pts/s)	75,000	75,000	70,000
Accuracy (mm)	0.012	0.009	0.006
Spacing distance (mm)	75	95	60
Depth/width FOV (mm)	3 × 65/3 × 65	60/-	15/-
Operating humidity range (%)	10–90%	10–90%	10–90%
Working temperature (°C)	10–40 °C	10–40 °C	10–40 °C

Table 3. Technical specification of products Metronor DUO and ATOS III Triple Scan.

Parameter	Metronor DUO	ATOS III Triple Scan
Price (EUR)	41,700	98,540
Weight (kg)	0.52	7.5
Dimensions (mm)	500 × 200 × 30	155 × 86 × 142
Measuring speed (pts/s)	35	75,000
Accuracy (mm)	0.025	0.01
Spacing distance (mm)	1500–15,000	490–2000
Depth FOV (Field of View) (mm)		230
Width FOV (Field of View) (mm)		250 × 250
Working temperature (°C)	10–45 °C	5–40 °C
Operating humidity range (%)	<90%	10–90%

The system can be supplemented with a touch probe working on a similar principle to the HandyPROBE and Metronor DUO, i.e., on the optical scanning of fixed reference points located on its body. The investment includes the already-mentioned KUKA mechanical arm and GOM Inspect Professional system control and measurement evaluation software.

2.1. Pairwise Comparison of Expert Evaluation Criteria

On the basis of personal recollection and interviews with experts, data were obtained to select the following criteria:

- Degree of education f_1;
- Practical experience with RE systems f_2;
- Theoretical knowledge in the field of quality inspection f_3;
- Economic knowledge f_4.

Practical experience is necessary so that the expert can evaluate the technical specifications of a particular product from its manual. In addition, they should know something about quality inspection because not all of the machines compared are able to measure to the required degree of accuracy [27,28]. A study of the firm's finances should ensure that there are sufficient funds to purchase a device with all the necessary gadgets. Finally, a reasonable level of education expresses the ability to carry out an analysis from the available documents and information (Table 4) [29,30].

Table 4. Pairwise criteria comparison for evaluation of experts.

Criterion	f_1	f_2	f_3	f_4
f_1	1	1/9	1/8	1/5
f_2	9	1	3	5
f_3	8	1/3	1	3
f_4	5	1/5	1/3	1

From the table above it is necessary to determine the characteristic polynomial by development of the following determinant:

$$\det(A_i - \lambda \cdot J) = 0 \tag{1}$$

where A_i is the pairwise comparison matrix and J is the identity matrix. Laplace's development of the determinant was used in the following calculation:

$$det|A| = \sum_{i=1}^{n} (-1)^{i+j} \cdot a_{ij} \cdot M_{ij}^A \tag{2}$$

where M_{ij}^A is the minor of a particular matrix $\left|S_{ij}^A\right|$.

From the calculated characteristic polynomial, we have identified the roots that are used to count the eigenvalue of the matrix [31], which are as follows:

$$\begin{gathered} \lambda_1 = 0.0186 + 0.28505\text{i} \\ \lambda_2 = 0.0186 - 0.28505\text{i} \\ \lambda_3 = 4.1746 \\ \lambda_4 = 0.1373 \end{gathered}$$

The eigenvalue of the matrix is then calculated according to the following formula:

$$max|\lambda_i| = \lambda_{max} = 4.1746 \tag{3}$$

Consistency index λ_{max} represents the condition for the acceptance of a reciprocal pairwise comparison matrix.

By further modification the consistency criterion can be calculated as follows:

$$\lambda_{max} - n \leq \alpha\left(\overline{\lambda}_{max}(n) - n\right) \tag{4}$$

Using the estimator $\overline{\lambda}_{max}$, i.e., applying the following approximation function:

$$\lambda_{max}(n) = 1.7699n - 4.3513 \tag{5}$$

The equation can be modified and thus obtain the consistency of our matrix as follows:

$$\begin{gathered} \lambda_{max} \leq n + \alpha(1.7699n - 4.3513) \\ 4.1746 \leq 4 + 0.1(1.7699 \times 4 - 4.3513) \\ 4.1746 \leq 4.2728 \end{gathered} \tag{6}$$

The calculation shows that the condition is valid and the criteria for pairwise comparison matrix of expert assessment can be accepted. The procedure substitutes the standard Saaty approach to computing the consistency ratio CI. For illustration purposes we also present his method of calculation below:

$$CR = CI/RI \leq 0.1.$$

where CR is the consistency ratio, CI is the consistency index, and RI is a random index. Let us say the coefficient is selected from the tables, its value for the matrix size $n = 4$ is 0.90 [32–37].

$$\begin{gathered} CI = \frac{\lambda_{max} - n}{n-1} = \frac{4.1746 - 4}{4-1} = 0.0582 \\ CR = \frac{0.0582}{0.90} \cong 0.0647 \leq 0.1 \end{gathered}$$

In the next step, the eigenvector of the matrix was determined by substituting eigenvalues to the system of equations of the form $(A - \lambda \cdot J)x = 0$, as follows :

$$\begin{vmatrix} 3.1746 & 1/9 & 1/8 & 1/5 \\ 9 & 3.1746 & 3 & 5 \\ 8 & 1/3 & 3.1746 & 3 \\ 5 & 1/5 & 1/3 & 3.1746 \end{vmatrix} \cdot \begin{vmatrix} x_1 \\ x_2 \\ x_3 \\ x_4 \end{vmatrix} = \begin{matrix} 0 \\ 0 \\ 0 \\ 0 \end{matrix}$$

The listed system of equations is a homogeneous system of linear equations and can have both zero and nonzero solutions (if the value of the matrix is less than n, it can have many linearly dependent solutions). To solve this system of linear equations one of the variables must be equal to 1. We have chosen x_1.

$$\begin{vmatrix} 3.1746 & 3 & 5 \\ 1/3 & 3.1746 & 3 \\ 1/5 & 1/3 & 3.1746 \end{vmatrix} \cdot \begin{vmatrix} x_2 \\ x_3 \\ x_4 \end{vmatrix} = \begin{matrix} -9 \\ -8 \\ -5 \end{matrix}$$

The solution of that system is the eigenvector of this matrix $x_1 = 1; x_2 = 14.6955; x_3 = 7.1342; x_4 = 3.2499$. We can calculate its normalized version as follows:

$$v_{kn_i} = \frac{v_{k_i}}{\sum_{j=1}^{k} v_{k_j}}$$

$$v_k = \begin{vmatrix} 1 \\ 14.6955 \\ 7.1342 \\ 3.2499 \end{vmatrix} \quad v_{kn} = \begin{vmatrix} 0.0383 \\ 0.5635 \\ 0.2736 \\ 0.1246 \end{vmatrix} \tag{7}$$

$$\sum v_{k_j} = 26.0796 \sum v_{kn_j} = 1$$

2.2. *Pairwise Comparison of Experts According to Individual Criteria*

The table below shows information about the pairwise comparison of experts according to selected individual criteria, based on the education of each expert.

Quality manager (a_1) and economist (a_3) have a university degree. Technician (a_2) has secondary technical education. The individual comparisons are listed in Table 5.

Table 5. Expert comparison by selected criteria.

Kr.	f_1				f_2				f_3				f_4			
	a_1	a_2	a_3	a_4	a_1	a_2	a_3	a_4	a_1	a_2	a_3	a_4	a_1	a_2	a_3	a_4
a_1	1	3	1	3	1	2	9	6	1	3	9	5	1	2	1/7	3
a_2	1/3	1	1/4	1/3	1/2	1	9	6	1/3	1	9	4	1/2	1	1/8	2
a_3	1	4	1	4	1/9	1/9	1	1/4	1/9	1/9	1	1/5	7	8	1	8
a_4	1/3	3	1/4	1	1/6	1/6	4	1	1/5	1/4	5	1	1/3	1/2	1/8	1

In the same way as in the case of the pairwise comparison of these criteria, we calculated all the necessary parameters for individual comparisons of variants according to the respective criteria from the matrices of pairwise comparisons mentioned above. The characteristic polynomials are as follows:

$$|A_{f1}| = \begin{vmatrix} 1-\lambda & 3 & 1 & 3 \\ 1/3 & 1-\lambda & 1/4 & 1/3 \\ 1 & 4 & 1-\lambda & 4 \\ 1/3 & 3 & 4 & 1-\lambda \end{vmatrix} \quad \Delta\left(\lambda_{Af1}\right) = \lambda^4 - 4\lambda^3 - \tfrac{17\lambda}{6} - \tfrac{1}{9}$$

$$|A_{f2}| = \begin{vmatrix} 1-\lambda & 2 & 9 & 6 \\ 1/2 & 1-\lambda & 9 & 6 \\ 1/9 & 1/9 & 1-\lambda & 1/4 \\ 1/6 & 1/6 & 4 & 1-\lambda \end{vmatrix} \quad \Delta\left(\lambda_{Af2}\right) = \lambda^4 - 4\lambda^3 - \tfrac{37\lambda}{12} - \tfrac{25}{48}$$

$$|A_{f3}| = \begin{vmatrix} 1-\lambda & 3 & 9 & 5 \\ 1/3 & 1-\lambda & 9 & 4 \\ 1/9 & 1/9 & 1-\lambda & 1/5 \\ 1/5 & 1/4 & 5 & 1-\lambda \end{vmatrix} \quad \Delta\left(\lambda_{Af4}\right) = \lambda^4 - 4\lambda^3 - \tfrac{99\lambda}{25} - \tfrac{616}{675}$$

$$|A_{f4}| = \begin{vmatrix} 1-\lambda & 2 & 1/7 & 3 \\ 1/2 & 1-\lambda & 1/8 & 2 \\ 7 & 8 & 1-\lambda & 8 \\ 1/3 & 1/2 & 1/8 & 1-\lambda \end{vmatrix} \quad \Delta\left(\lambda_{Af4}\right) = \lambda^4 - 4\lambda^3 - \tfrac{107\lambda}{56} - \tfrac{13}{112}$$

From the characteristic polynomial $\Delta(\lambda_{Afi})$ of individual matrices A_{fi} we calculated their roots. According to Equation (4) we determined the parameters of lambda, and according to Equation (6) we assessed the consistency of the matrices, as follows:

$$A_{f1} = \begin{vmatrix} \lambda_3 = 4.1649 \\ \lambda_1 = -0.0629 + 0.8233i \\ \lambda_2 = -0.0629 - 0.8233i \\ \lambda_4 = -0.0391 \end{vmatrix} \Rightarrow \lambda_{max} = 4.1649 \Rightarrow 4.1649 \leq 4.2728$$

$$A_{f2} = \begin{vmatrix} \lambda_3 = 4.1833 \\ \lambda_1 = -0.0101 + 0.8233i \\ \lambda_2 = -0.0101 - 0.8233i \\ \lambda_4 = -0.1631 \end{vmatrix} \Rightarrow \lambda_{max} = 4.1833 \Rightarrow 4.1833 \leq 4.2728$$

$$A_{f3} = \begin{vmatrix} \lambda_3 = 4.2330 \\ \lambda_1 = -0.0069 + 0.9916i \\ \lambda_2 = -0.0069 - 0.9916i \\ \lambda_4 = -0.2192 \end{vmatrix} \Rightarrow \lambda_{max} = 4.233 \Rightarrow 4.233 \leq 4.2728$$

$$A_{f4} = \begin{vmatrix} \lambda_3 = 4.1145 \\ \lambda_1 = -0.0271 + 0.6835i \\ \lambda_2 = -0.0271 - 0.6835i \\ \lambda_4 = -0.0603 \end{vmatrix} \Rightarrow \lambda_{max} = 4.1145 \Rightarrow 4.233 \leq 4.2728$$

The consistency condition is met in the case of all the comparisons, so we can accept the matrices. The eigenvalues of the matrices and the eigenvalues of vectors of the matrices are as follows:

$$v_{k1} = \begin{vmatrix} 1 \\ 0.2419 \\ 1.1604 \\ 0.4263 \end{vmatrix} \quad v_{kn1} = \begin{vmatrix} 0.3535 \\ 0.0855 \\ 0.4103 \\ 0.1507 \end{vmatrix}$$

$$v_{k2} = \begin{vmatrix} 1 \\ 0.7106 \\ 0.074 \\ 0.1826 \end{vmatrix} \quad v_{kn2} = \begin{vmatrix} 0.5083 \\ 0.3612 \\ 0.0376 \\ 0.0928 \end{vmatrix}$$

$$v_{k3} = \begin{vmatrix} 1 \\ 0.5393 \\ 0.0656 \\ 0.205 \end{vmatrix} \quad v_{kn3} = \begin{vmatrix} 0.5525 \\ 0.298 \\ 0.0362 \\ 0.1133 \end{vmatrix}$$

$$v_{k4} = \begin{vmatrix} 1 \\ 0.6117 \\ 4.8453 \\ 0.3997 \end{vmatrix} \quad v_{kn4} = \begin{vmatrix} 0.1458 \\ 0.0892 \\ 0.7067 \\ 0.0583 \end{vmatrix}$$

The table below provides an overview of the importance of individual experts involved in the investment selection. The values of the scalar product of the weight vector and the specific vectors representing the experts are displayed in the weighted sum line. As can be observed, the decisions of the quality manager will have the greatest weight at this level of the hierarchical structure. He is followed by his technician, economist, and engineer [38,39].

3. Results

Based on the information available, all the experts have been compared against each other according to particular criteria. In this way, the weight of each expert has been calculated. The results are recorded in the table below (Table 6). These values will subsequently be used in the product selection process.

Table 6. Weights of particular experts.

Criterion	Weight	Quality Manager	Technician	Economist	Engineer
Education degree	0.0383	0.3535	0.0855	0.4103	0.1507
Practical experiences with RE systems	0.5635	0.5083	0.3612	0.0376	0.0928
Theoretical knowledge from quality inspection	0.2736	0.5525	0.298	0.0362	0.1133
Economic knowledge	0.1246	0.1458	0.0892	0.7067	0.0583
Weighted sum		0.4693	0.2995	0.1349	0.0963

In the same way, the weight of each selected criterion of the scanners has been calculated. Pairwise comparison by members of the evaluation team at this level of hierarchy structure should reflect their personal preferences, according to which they will give priority to a specific product. The results are shown in the following table (Table 7).

Table 7. Weight of product assessment criteria.

Expert	Weight	f_1	f_2	f_3	f_4	f_5	f_6	f_7
Quality manager	0.4693	0.0228	0.4286	0.1083	0.1293	0.1582	0.1170	0.0358
Technician	0.2995	0.0232	0.4391	0.1958	0.0718	0.0597	0.1299	0.0805
Economist	0.1349	0.3375	0.2646	0.0352	0.0667	0.0887	0.1499	0.0574
Engineer	0.0963	0.0245	0.4574	0.1734	0.0943	0.0806	0.1170	0.0528
Weighted sum:		0.0655	0.4124	0.1309	0.1003	0.1118	0.1253	0.0537

The last step in the analysis is to process the matrices of pairwise comparisons, in which every product is assessed according to each criterion for product assessment

(Table 8). The considered variants are as follows: MetraSCAN 70 (a_1), MetraSCAN 70-R (a_2), HandyPROBE (a_3), Nikon XC65Dx (a_4), Nikon LC60Dx (a_5), Nikon LC15Dx (a_6), Metronor DUO (a_7), and ATOS Triple Scan (a_8).

Table 8. Final product order.

Criterion	Weight	a_1	a_2	a_3	a_4	a_5	a_6	a_7	a_8
Price (f_1)	0.0655	0.0894	0.0204	0.2930	0.0497	0.1252	0.1252	0.2473	0.0497
Accuracy (f_2)	0.4124	0.0266	0.0198	0.0526	0.1292	0.2216	0.3488	0.0640	0.1375
Portability (f_3)	0.1309	0.2288	0.0326	0.3073	0.0310	0.0310	0.0310	0.3073	0.0310
Sensed area (f_4)	0.1003	0.1163	0.1163	0.0156	0.2009	0.0910	0.0584	0.0157	0.3859
Depth of field (f_5)	0.1118	0.0339	0.0339	0.2928	0.0516	0.0699	0.0204	0.2784	0.2190
Sensing rate (f_6)	0.1253	0.1305	0.1174	0.0175	0.1714	0.0800	0.0567	0.0165	0.4100
Possibility to move with sensed object & ease of implement. (f_7)	0.0537	0.2082	0.2082	0.2082	0.0334	0.0334	0.0334	0.2416	0.0334
Weighted sum:		0.0898	0.0551	0.1288	0.1098	0.1324	0.1731	0.1306	0.1804
Order:		7	8	5	6	3	2	4	1

As can be observed in the table above, the ATOS Triple Scan corresponds best to all the stated criteria, and slightly worse is the Nikon LC15Dx. In this case, the best solution is not the cheapest, for these two machines are more expensive than the others. Using this method, and by engaging staff in the process, a company can prevent many mistakes before buying and implementing any equipment under consideration. Before choosing one of the solutions, the firm can perform an additional analysis of the profitability of an investment, so that they can select the best option.

4. Discussion

The figure below (Figure 3) shows the whole four-level hierarchical structure, in which the calculated weights of the evaluation team, the criteria for the equipment, and, finally, on the fourth level, the weight determining the final order of alternatives can be observed.

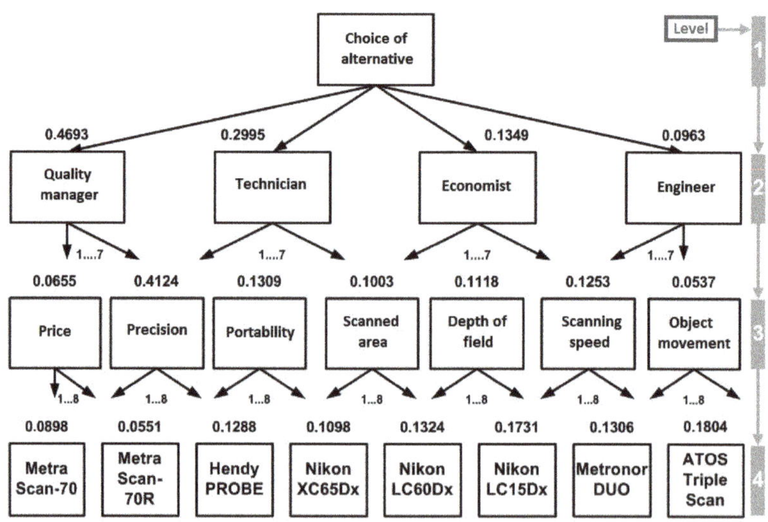

Figure 3. Diagram of multilevel hierarchical structure with values of individual weights.

Due to the fact that the AHP method requires relatively demanding mathematical calculations, and the matrices themselves are larger, a function (see Algorithm 1) in the MATLAB program was used to perform the calculations. The function itself verifies the inverse axiom; let us say that the reciprocity rule $rji = 1/rij$, which, if not met, will print an error message. In the next step, it checks the consistency of the matrix (if it is not consistent, it returns an error message), and, finally, it calculates the required vectors.

Algorithm 1. MATLAB program

```
[x,xn] = ahp(A), Where x is a proprietary vector of matrix A, xn is a proprietary vector of matrix A
(x/sum x) and alfa is a level of consistence
function [x,xn] = ahp(A,alfa)
% Check that the matrix is entered correctly
for i = 1:size(A,1)
for j = 1:size(A,1)
if i~= j
if A(j,i) == A(i,j) & A(j,i) ~= 1;
str=['The matrix is misspelled: element a_' ...
,num2str(i),num2str(j),' ','=',' ' ...
,'a_',num2str(j),num2str(i)];
disp(str)
return;
elseif A(j,i) ~= 1/A(i,j)
str = ['The matrix is misspelled: element a_' ...
,num2str(i),num2str(j),' ','a',' ' ...
,'a_',num2str(j),num2str(i),' ', ...
' not in the desired shape a_ij = 1/a_ji'];
disp(str)
return;
end
end
end
% Check of matrix A consistency.
if eigs(A,1)<= size(A,1) + alfa*(1.7699*size(A) − 4.3513)
% Calculation of eigenvector of matrix (x) and the eigenvalue vector
% matrix (xn) according formula (A − lambda*J)x = 0
x = (A − eigs(A,1)*eye(size(A,1)));
x = [1;inv(x(2:size(A),2:size(A)))*(−A(2:size(A),1))];
xn = x/sum(x);
else str = ['Matrix on level alfa=',num2str(alfa),' ',' is not consistent' ];
disp(str)
end
```

5. Conclusions

The presented information shows the way in which it is possible to proceed with the selection of a scanning device for reverse engineering for a company. The overall methodology is presented, from the selection of important criteria, through their evaluation, to the selection of the most appropriate technical solution. It is a complex, but demanding, methodology that can objectify selection. The article presents the whole procedure applied to a specific example, with the weight of individual criteria. A very effective AHP pairwise comparison methodology is used. The paper presents a novel application of the AHP method, using, as an example, the selection of a reverse engineering scanning device, and introduces a new perspective in this field. Among the many benefits it brings are enhanced effectiveness and cost saving. This evaluation process was performed in the company under real conditions and verified in real terms.

Author Contributions: Conceptualization, J.B. and M.M.; methodology, J.B.; software, P.K.; validation, L'.Š. and J.B.; formal analysis, J.B.; investigation, J.B.; resources, J.B. and P.K.; data curation, J.B.;

writing—original draft preparation, J.B. and M.M.; writing—review and editing, J.B.; visualization, J.B.; supervision, J.B. and P.K.; project administration, J.B.; funding acquisition, J.B. All authors have read and agreed to the published version of the manuscript.

Funding: The paper is a part of the research conducted within the project APVV-18-0527 "Development and optimization of additive manufacturing technology and design of device for production of components with optimized strength and production costs" funded by the Slovak Research and Development Agency. This paper was completed in association with the project "Innovative and additive manufacturing technology—new technological solutions for 3D printing of metals and composite materials", reg. no. CZ.02.1.01/0.0/0.0/17_049/0008407 financed by Structural Funds of the European Union and project. The paper is a part of the research conducted within the project VEGA 1/0665/21 "Research and optimization of technological parameters of progressive additive manufacturing of effective protective equipment against COVID-19" funded by the Ministry of Education of Slovak Republic and to the Slovak Academy of Sciences.

Institutional Review Board Statement: Not applicable.

Informed Consent Statement: Not applicable.

Data Availability Statement: The data are available at Slovak University of Technology in Bratislava.

Conflicts of Interest: The authors declare no conflict of interest.

References

1. Dúbravčík, M.; Kender, S. Application of Reverse Engineering Techniques in Mechanics System Services. *Procedia Eng.* **2012**, *48*, 96–104. [CrossRef]
2. Buonamici, F.; Carfagni, M.; Furferi, R.; Governi, L.; Lapini, A.; Volpe, Y. Reverse engineering of mechanical parts: A template-based approach. *J. Comput. Des. Eng.* **2017**, *5*, 145–159. [CrossRef]
3. Moon, S.; Ko, K. A point projection approach for improving the accuracy of the multilevel B-spline approximation. *J. Comput. Des. Eng.* **2017**, *5*, 173–179. [CrossRef]
4. Rysiński, J.; Wrobel, I. Diagnostics of machine parts by means of reverse engineering procedures. *Adv. Mech. Eng.* **2015**, *7*, 1687814015584543. [CrossRef]
5. Docchio, F.; Sansoni, G.; Trebeschi, M. Inspection, 3D modelling, and rapid prototyping of cultural heritage by means of a 3D optical digitiser. *Opt. Methods Arts Archaeol.* **2005**, *5857*, 58570D. [CrossRef]
6. Pandilov, Z.; Betim, S.; Dejan, S. Reverse engineering an effective tool for design and development of mechanical parts. *Acta Tech. Corviniensis–Bull. Eng.* **2018**, *11*, 113–118.
7. Atanasova-Pacemska, T.P.; Lapevski, M.; Timovski, R. Analytical hierarchical process (ahp) method application in the process of selection and evaluation. In Proceedings of the UNITECH—International Scientific Conference, Gabrovo, Bulgaria, 21–22 November 2014; Volume 14.
8. Buonamici, F.; Carfagni, M.; Furferi, R.; Governi, L.; Lapini, A.; Volpe, Y. Reverse engineering modeling methods and tools: A survey. *Comput. Des. Appl.* **2017**, *15*, 443–464. [CrossRef]
9. Eldad, E. *Reversing: Secrets of Reverse Engineering*; Wiley Publishing, Inc.: Indianapolis, ID, USA, 2015; ISBN 0-7645-7481-7.
10. Thompson, W.; Owen, J.; Germain, H.D.S.; Stark, S.; Henderson, T. Feature-based reverse engineering of mechanical parts. *IEEE Trans. Robot. Autom.* **1999**, *15*, 57–66. [CrossRef]
11. Kumar, A.; Jain, P.; Pathak, P. Conception of Part Reconstruction: Integration of Non-Contact Scanning and Rapid Prototyping. In Proceedings of the Conference AM-2014, Bengaluru, India, 1–2 September 2014.
12. Prochazkova, J.; Procházka, D.; Landa, J. Sharp Feature Detection as a Useful Tool in Smart Manufacturing. *ISPRS Int. J. Geo-Inf.* **2020**, *9*, 422. [CrossRef]
13. Miądlicki, K.; Jasiewicz, M.; Gołaszewski, M.; Królikowski, M.; Powałka, B. Remanufacturing System with Chatter Suppression for CNC Turning. *Sensors* **2020**, *20*, 5070. [CrossRef]
14. Raut, L.; Barai, G.; Shete, S. Design and development of a component by reverse engineering. *Int. J. Res. Eng.Technol.* **2015**, *4*, 539–546.
15. Ivana, R.; Zuzana, M. The Analysis of AHP method and its potential use in logistics. *Acta Montanistica Slovaca* **2009**, *14*, 103–112.
16. Gospodarek, M.; Rybarczyk, P.; Szulczyński, B.; Gębicki, J. Comparative Evaluation of Selected Biological Methods for the Removal of Hydrophilic and Hydrophobic Odorous VOCs from Air. *Processes* **2019**, *7*, 187. [CrossRef]
17. Ruiz-Ramos, J.; Marino, A.; Boardman, C.; Suarez, J. Continuous Forest Monitoring Using Cumulative Sums of Sentinel-1 Timeseries. *Remote Sens.* **2020**, *12*, 3061. [CrossRef]
18. Koczkodaj, W.W.; Szybowski, J.; Wajch, E. Inconsistency indicator maps on groups for pairwise comparisons. *Int. J. Approx. Reason.* **2016**, *69*, 81–90. [CrossRef]
19. Leal, J.E. AHP-express: A simplified version of the analytical hierarchy process method. *MethodsX* **2019**, *7*, 100748. [CrossRef] [PubMed]

20. Freerk, A.L. *Multi-Criteria Decision Analysis via Ratio and Difference Judgement*; Applied Optimization; Springer: Boston, MA, USA, 1999. [CrossRef]
21. Viana, V.R.; Ipma-B, P. *Using the Analytic Hierarchy Process (AHP) to Select and Prioritize Projects in a Portfolio*; PMI Global Congress: Washington, DC, USA, 2010; pp. 1–12.
22. Podvezko, V.; Mitkus, S.; Trinkūniene, E. Complex evaluation of contracts for construction. *J. Civ. Eng. Manag.* **2010**, *16*, 287–297. [CrossRef]
23. Kim, S.; Kim, B. A Decision-Making Model for Adopting Al-Generated News Articles: Preliminary Results. *Sustainability* **2020**, *12*, 7418. [CrossRef]
24. Kristbaum, J.P.; Ciarallo, F.W. Strategic Decision Facilitation: Supporting Critical Assumptions of the Human in Empirical Modeling of Pairwise Value Comparisons. *Systems* **2020**, *8*, 30. [CrossRef]
25. Derrek, P.H.; Neda, J.; Sarah, E.; Medland, P.M. *Thompson, Chapter Nine-Continuous Inflation Analysis: A Threshold-Free Method to Estimate Genetic Overlap and Boost Power in Imaging Genetics*; Adrian, V., Dalca, N., Batmanghelich, K., Li, S., Mert, R.S., Eds.; Imaging Genetics; Academic Press: Cambridge, MA, USA, 2018; pp. 147–162, ISBN 9780128139684. [CrossRef]
26. Nordstokke, D.; Stelnicki, A.M. Pairwise Comparisons. In *Encyclopedia of Quality of Life and Well Being Research*; Michalos, A.C., Ed.; Springer: Dordrecht, The Netherlands, 2014. [CrossRef]
27. Sébastien, R.; Laroche, F. Durupt Alexandre, Bernard Alain, Knowledge Based Reverse Engineering Methodology. In Proceedings of the ASME 2012 11th Biennial Conference on Engineering Systems Design and Analysis, ESDA 2012, Nantes, France, 2–4 July 2012. [CrossRef]
28. Alexander, M. *Manažérstvo Kvality: História, Koncepty, Metódy*, 1st ed.; Epos: 2006; p. 752, ISBN 80-8057-656-4. Available online: https://books.google.com.hk/books/about/Mana%C5%BE%C3%A9rstvo_kvality.html?id=85u2AAAACAAJ&redir_esc=y (accessed on 14 December 2021).
29. Creath, K.; Wyant, J.C. Moiré and Fringe Projection Techniques. In *Optical Shop Testing*, 2nd ed.; Malacara, D., Ed.; John Wiley & Sons, Inc.: Hoboken, NJ, USA, 1992; pp. 653–685.
30. Zalai, K.; Kalafutová, E.; Šnircová, J. *Financial and Economic Analysis of the Company*; Sprint: Bratislava, Slovensko, 2002; p. 305, ISBN 80-88848-94-6.
31. Várady, T.; Martin, R.R.; Cox, J. Reverse engineering of geometric models—An introduction. *Comput. Des.* **1997**, *29*, 255–268. [CrossRef]
32. Svetlík, J.; Demeč, P. Methods of Identifying the Workspace of Modular Serial Kinematic Structures. *Appl. Mech. Mater.* **2013**, *309*, 75–79. [CrossRef]
33. Alonso, J.A. Lamata Teresa, Consistency in the analytic hierarchy process: A new approach. *Int. J. Uncertain. Fuzziness Knowl.-Based Syst.* **2006**, *14*, 445–459. Available online: http://hera.ugr.es/doi/16515833.pdf (accessed on 10 November 2021).
34. Raja, V.; Fernandes, K.J. *Fernandes, Reverse Engineering—An Industrial Perspective*; Springer: London, UK, 2008; ISBN 978-1-84628-855-5. [CrossRef]
35. Park, J.; DeSouza, G.N. 3-D Modeling of Real-World Objects Using Range and Intensity Images. In *Machine Learning and Robot Perception*; Springer: Berlin/Heidelberg, Germany, 2005; pp. 203–264. [CrossRef]
36. Rocchini, C.; Cignoni, P.; Montani, C.; Pingi, P.; Scopigno, R. A low cost 3D scanner based on structured light. *Comput. Graph. Forum* **2001**, *20*, 299–308. [CrossRef]
37. Boehler, W.; Marbs, A. The potential of non-contact close range laser scanners for cultural heritage recording. *Int. Arch. Photogramm. Remote Sens. Spat. Inf.Sci.* **2002**, *34*, 430–436.
38. Beniak, J.; Križan, P.; Matus, M.; Svatek, M. Ecological PLA plastic used for FDM rapid prototyping technology. In Proceedings of the International Multidisciplinary Scientific GeoConference Surveying Geology and Mining Ecology Management, SGEM, Albena, Bulgaria, 18–24 June 2015; Volume 1, pp. 117–123. [CrossRef]
39. Svetlik, J.; Demeč, P. Methods of Identifying the Workspace of Modular Serial Kinematic Structures. In Proceedings of the Applied Mechanics and Materials: CECOL 2012: 3rd Central European Conference on Logistics, Trnava, Slovakia, 28–30 November 2012; Volume 309, pp. 75–79, ISBN 978-303785636-9.

MDPI
St. Alban-Anlage 66
4052 Basel
Switzerland
Tel. +41 61 683 77 34
Fax +41 61 302 89 18
www.mdpi.com

Applied Sciences Editorial Office
E-mail: applsci@mdpi.com
www.mdpi.com/journal/applsci

CPSIA information can be obtained
at www.ICGtesting.com
Printed in the USA
LVHW070832230223
740229LV00010B/482

9 783036 562810